AF238973

Marcus Baum

**Simultaneous Tracking and Shape
Estimation of Extended Objects**

Karlsruhe Series on Intelligent Sensor-Actuator-Systems

Volume 13

ISAS | Karlsruhe Institute of Technology
        Intelligent Sensor-Actuator-Systems Laboratory

Edited by Prof. Dr.-Ing. Uwe D. Hanebeck

Simultaneous Tracking and Shape Estimation of Extended Objects

by
Marcus Baum

Dissertation, Karlsruher Institut für Technologie (KIT)
Fakultät für Informatik, 2013

Impressum

 Scientific
Publishing

Karlsruher Institut für Technologie (KIT)
KIT Scientific Publishing
Straße am Forum 2
D-76131 Karlsruhe

KIT Scientific Publishing is a registered trademark of Karlsruhe
Institute of Technology. Reprint using the book cover is not allowed.

www.ksp.kit.edu

Print on Demand 2013

ISSN 1867-3813
ISBN 978-3-7315-0078-0

Simultaneous Tracking and Shape Estimation of Extended Objects

zur Erlangung des akademischen Grades eines

Doktors der Ingenieurwissenschaften

von der Fakultät für Informatik
des Karlsruher Instituts für Technologie (KIT)

genehmigte

Dissertation

von

Marcus Baum

aus Heidelberg

Tag der mündlichen Prüfung: 18.01.2013

Erster Gutachter: Prof. Dr.-Ing. Uwe D. Hanebeck

Zweiter Gutachter: Prof. Peter K. Willett, PhD

Acknowledgements

This thesis emerged from my time as a research assistant at the Intelligent Sensor Actuator (ISAS) lab at the Karlsruhe Institute of Technology (KIT). During my work on this thesis, I was supported by many people in various ways. At this point, I would like to appreciate especially those mentioned below.

First and foremost, my deepest gratitude goes to Uwe D. Hanebeck for advising me and giving me the opportunity to work in his lab. Without his continuous support and helpful guidance this thesis would not have come into existence. Furthermore, I am very grateful to Peter K. Willett for co-advising me and inviting me to the University of Connecticut. Due to his exceptional hospitality and many inspiring discussions, my research stay in Connecticut was an unforgettable and rewarding experience.

A special thank goes to the current and former research assistants at ISAS for their collaboration, friendship, and the great time we had together. In particular, I would like to name my co-authors Frederik Beutler, Florian Faion, Vesa Klumpp, Benjamin Noack, Marc Reinhardt, and Patrick Ruoff. I am also grateful to the numerous students that worked together with me on different research projects, especially Kai Bouché, Dominik Itte, Robin Sandkühler, and Antonio Zea. Furthermore, I am very much obliged to the secretaries and the technical staff at ISAS for their administrative and technical help in uncountable ways. Thanks go also to the Karlsruhe House of Young Scientists (KHYS) for supporting my research stay at the University of Connecticut.

I would like to extend my appreciation to all my family members who supported me along the way. Particular thanks are due to my parents, sisters, and nieces for their loving care and understanding. Last but not least, I would like to thank my beloved Mira for her everlasting patience and belief in my ambitions.

Karlsruhe, May 2013   Marcus Baum

Contents

Zusammenfassung

Die vorliegende Arbeit beschäftigt sich mit dem Verfolgen (Tracking) eines mobilen Objekts basierend auf verrauschten Sensormessungen. Klassische Trackingalgorithmen nehmen in der Regel an, dass das mobile Objekt aufgrund von starkem Sensorrauschen als ein Punkt ohne Ausdehnung modelliert werden kann. Bei modernen, hochauflösenden Sensoren ist diese Annahme jedoch nicht gerechtfertigt und es muss berücksichtigt werden, dass Messungen von verschiedenen, räumlich verteilten Messquellen auf dem Objekt stammen können. Da die räumliche Ausdehnung des mobilen Objekts nicht bekannt ist und sich im Laufe der Zeit ändern kann, muss simultan zum Tracking eine Formschätzung durchgeführt werden. Hierbei ergeben sich zwei grundlegende Herausforderungen: i) Aufgrund der unbekannten Form handelt es sich um ein hochdimensionales, nichtlineares Schätzproblem, dessen Komplexität mit Standardmethoden nicht beherrschbar ist. ii) Da neben der Form auch die Beschaffenheit des Objekts unbekannt ist, liegen keine verlässlichen Informationen darüber vor, wo auf dem Objekt Messquellen sein könnten. Der Trackingalgorithmus sollte sich daher für unterschiedlich beschaffene Objekttypen eignen und robust gegenüber möglichen Modellierungsfehlern sein.

Fitting von Kreisen und Ellipsen
Wird das ausgedehnte Objekt als eine geschlossene Kurve, z. B. ein Kreis, modelliert, kann die Formschätzung als ein dynamisches Fitting-Problem aufgefasst werden. Diesbezüglich wird in dieser Arbeit ein neues Verfahren zum Fitting von Kreisen und Ellipsen vorgestellt. Die Grundidee ist, eine explizite Messabbildung mit multiplikativem Messrauschen aufzustellen und eine statistische Linearisierung durchzuführen, um ein effizientes, rekursives Messupdate mit einem Gaußschen Filter zu ermöglichen. Insbesondere bei starkem Messrauschen ist dieser neue Ansatz Standardmethoden überlegen.

Random Hypersurface Model
Desweiteren werden ausgedehnte Objekte behandelt, bei denen die Messungen von einer Fläche, z. B. einer Kreisscheibe, stammen. Hierzu wird

ein neues Konzept namens Random Hypersurface Model (RHM) zur Beschreibung von Messquellen auf einer Fläche, d. h. im Inneren einer geschlossenen Kurve, vorgestellt. Ein RHM nimmt an, dass jede Messquelle auf einer skalierten Version des Flächenrands liegt, wobei der Skalierungsfaktor durch eine Wahrscheinlichkeitsdichte charakterisiert wird. Auf diese Weise wird das Modellieren einer Fläche auf das Modellieren einer Kurve reduziert. Basierend auf dem Konzept von RHMs und den neuentwickelten Techniken zum Fitting von Kurven werden effiziente Gaußsche Filter für Grundformen, z. B. Ellipsen, aber auch für beliebige sternkonvexe Formen hergeleitet. Hiermit wird gezeigt, dass die Komplexität des Schätzproblems beherrschbar ist und eine detailreiche Form aus verrauschten Messungen effizient geschätzt werden kann. Diese detailreiche Formschätzung ist von großem Nutzen für das gesamte Trackingsystem, da sie z. B. zur Klassifikation des Objekts verwendet werden kann.

Mengenbasiertes Ausdehnungsmodell
Zur Vermeidung von Modellierungsfehlern bei den Messquellen wird in einem weiteren Ansatz ein mengentheoretisches Ausdehnungsmodel vorgeschlagen, d. h. es werden keine statistischen Annahmen über die Positionen der Messquellen auf dem Objekt gemacht. Dieser Ansatz führt auf einen kombiniert mengenbasierten und stochastischen Schätzer, für den spezielle Approximationstechniken entwickelt werden. Aufgrund des mengentheoretischen Modells ist der Schätzer robust gegenüber systematischen Fehlern in den Positionen der Messquellen. Jedoch ist es nicht möglich, die Form des Objekts allein aus Punktmessungen zu schätzen. Der mengenbasierte Ansatz bietet sich demnach an, sofern kein Wissen über die Verteilung der Messquellen vorliegt und die Form des Objekts bekannt ist oder aus anderen Informationsquellen geschätzt werden kann (z. B. über die Anzahl der Messungen). Es wird gezeigt, dass dieser Ansatz robustere Schätzergebnisse liefert als ein rein stochastisches Verfahren, wenn das stochastische Modell des Objekts nicht mit dem wahren übereinstimmt.

Experimente
Als Beispielanwendung wird das Tracking eines mobilen Objekts auf einem Tisch betrachtet. Hierfür wird ein Sensor, der Farb- und Tiefenbilder aus der Vogelperspektive liefert, verwendet, um sich bewegende Punkte auf dem Tisch zu detektieren. Die extrahierten Punkte stammen von der Oberfläche des mobilen Objekts und können somit zur Formschätzung verwendet werden. Anhand dieses Experiments werden die Vorteile der entwickelten Methoden im Vergleich mit Standardmethoden dargelegt.

Summary

This thesis is devoted to the tracking of a mobile object based on noisy sensor measurements. Traditional tracking algorithms assume that the mobile object can be modeled as a point without any extent due to high sensor noise. However, this assumption is not reasonable for modern high-resolution sensor devices because an object may give rise to multiple measurements from spatially distributed measurement sources on the object. As the spatial extent of the object is unknown and may vary over time, it becomes necessary to simultaneously track the object and estimate its shape. For this purpose, the following two basic challenges arise: i) As the shape is unknown and to be estimated, extended object tracking is a high-dimensional, nonlinear estimation problem whose complexity cannot be handled with standard methods. ii) The characteristics of the object's surface are unknown as well, so that no reliable information about possible measurement sources is available. Hence, it must be ensured that the tracking algorithm is robust to modeling errors caused by different kinds of objects.

Fitting of Circles and Ellipses
When the extended object is modeled as a closed curve, e.g., a circle, the shape estimation can be interpreted as a dynamic curve fitting problem. In this context, a novel method for fitting circles and ellipses to noisy point measurements is proposed. The basic idea is to reformulate the estimation problem as an explicit measurement equation that is corrupted with multiplicative noise. Statistical linearization of this measurement equation allows for an efficient, recursive measurement update with a Gaussian filter. In particular for large measurement noise, this novel approach is able to significantly outperform standard methods.

Random Hypersurface Model
Next, extended objects that are modeled as regions such as circular discs are treated. For this purpose, a novel concept called Random Hypersurface Model (RHM) for specifying measurement sources in a region, i.e., the interior of a closed curve, is proposed. An RHM assumes that each measurement source lies on a scaled version of the region boundary, where

the scaling factor is characterized by a probability density function. In this vein, the modeling of a region is reduced to the modeling of a curve, i.e., the boundary. Based on the concept of an RHM and the developed curve fitting techniques, efficient Gaussian filters for basic shapes such as ellipses and star-convex shapes are derived. By this means, it is shown that extended object tracking is tractable and it is possible to extract detailed shape information based on sequentially arriving noisy measurements. Such a detailed shape estimate is a significant benefit for the entire tracking system, e.g., it can be used for classification.

Set-Theoretic Extent Model
In order to cope with the lack of statistical knowledge about the measurement sources, a further novel approach is proposed that uses a set-theoretic extent model, i.e., no statistical assumptions are made about possible measurement sources on the object. This approach leads to a combined set-theoretic and statistical estimator for which special approximation techniques are developed. Due to the set-theoretic model for the measurement sources, the estimator is robust to systematic modeling errors in the locations of the measurement sources. However, it is not possible to estimate the shape of the object solely with point measurements. Hence, the set-theoretic approach is suitable in case no statistical assumptions on the measurement sources are justified and the shape parameters are known or can be estimated from other information sources (e.g., the number of received measurements). It turns out that the set-theoretic approach is able to give more robust estimates than plain stochastic methods, especially when the stochastic object model significantly differs from the true object.

Experiments
As an example application, the tracking of a ground moving mobile object on a table is considered. For this purpose, a sensor that supplies RGB and depth images from a bird's eye view is employed for detecting moving points on the table. The extracted points originate from the surface of the moving object and hence, can be used for estimating the object's shape. With the help of this experiment, the advantages of the developed methods are highlighted with respect to standard methods.

Notation

Glossary

EKF	*Extended Kalman Filter*
UKF	*Unscented Kalman Filter*
LRKF	*Linear Regression Kalman Filter*
RMSE	*Root-Mean-Square Error*
RHM	*Random Hypersurface Model*
SSI filter	*Statistical and Set-theoretic Information (SSI) Filter*
UKF-EL-RHM	*UKF implementation of an elliptic RHM*
AMC-SC-RHM	*Analytic implementation of a star-convex RHM*
UKF-SC-RHM	*UKF implementation of a star-convex RHM*
PF-SC-RHM	*Particle filter implementation of a star-convex RHM*
Naïve-PF	*Naïve particle filter implementation for extended objects*

General Conventions

x, y	Scalar
$\underline{x}$, $\underline{y}$	Vector
$\underline{x}^T$	Transpose of $\underline{x}$
$\underline{\boldsymbol{x}}$	Random vector
$\mathbf{A}$	Matrix
$\mathbf{I}_n$	Identity matrix of dimension n
$\mathbb{N}$	Natural numbers
$\mathbb{R}$	Real numbers
$\mathrm{E}\{\underline{\boldsymbol{x}}\} = \underline{\mu}^x$	Mean of $\underline{\boldsymbol{x}}$
$\mathrm{Cov}\{\underline{\boldsymbol{x}}\} = \Sigma^x$	Covariance matrix of $\underline{\boldsymbol{x}}$
$\mathcal{N}(\underline{x} - \underline{\mu}^x, \Sigma^x)$	Gaussian distribution with mean $\underline{\mu}^x$ and covariance matrix Σ^x
$f(\underline{x})$	Probability density function of $\underline{\boldsymbol{x}}$
$F(\underline{x})$	Cumulative probability distribution of $\underline{\boldsymbol{x}}$

Conventions for Extended Object Tracking

$\underline{x}_k$, $\underline{\boldsymbol{x}}_k$	State vector at time step k
$\underline{p}_k$, $\underline{\boldsymbol{p}}_k$	Shape parameters at time step k
$\underline{m}_k$, $\underline{\boldsymbol{m}}_k$	Center of the extended object at time step k
$\mathcal{S}(\underline{p}_k)$	Shape specified by $\underline{p}_k$
$\mathcal{M}(\underline{x}_k)$	Set of measurement sources specified by $\underline{x}_k$
$\partial\mathcal{S}(\underline{p}_k)$	Boundary of $\mathcal{S}(\underline{p}_k)$
$\underline{z}_{k,l}$, $\underline{\boldsymbol{z}}_{k,l}$	l-th measurement source at time step k
$\underline{\hat{y}}_{k,l}$, $\underline{\boldsymbol{y}}_{k,l}$	l-th measurement at time step k
$\mathcal{y}_k$, $\hat{\mathcal{y}}_k$	All received measurements at time step k
$\underline{\boldsymbol{v}}_{k,l}$	Measurement noise for l-th measurement at time step k
$\underline{\boldsymbol{w}}_k$	System noise at time step k
$\mathbf{K}\left(\underline{\xi}\right)$	Circle with parameter vector $\underline{\xi}$
$\mathbf{C}\left(\underline{\eta}\right)$	Cone with parameter vector $\underline{\eta}$
$\mathbf{H}\left(\underline{\zeta}\right)$	Hyperboloid with parameter vector $\underline{\zeta}$
$\Delta_{k,l}$	Solution set for time step k having incorporated the first l measurements
$\Theta_{k,l}$	Measurement solution set for time step k having incorporated the first l measurements

CHAPTER 1

Introduction

Contents

Today, life is hard to imagine without technical devices that are equipped with sensors. Sensors are the senses for technical devices and similar to human senses, a main purpose of sensors is to answer the fundamental questions "Where am I?" and "Where are the others?". From a technical point of view, the answer to these questions leads to the term *localization*, which denotes the process of determining the pose, i.e., the position and orientation, of an object. The continuous localization of a mobile object over time is in general referred to as *object tracking*.

Essentially, two different settings for object tracking can be distinguished: When the mobile object is actively involved in the localization process, e.g., when sensors are mounted on the mobile object, we talk about *cooperative tracking*. For example, the tracking of a mobile phone with the help of its GPS sensor is a cooperative tracking problem. In contrast, *non-cooperative tracking* denotes the process of localizing an object from remote sensors in the environment. A classical example for non-cooperative tracking is air traffic surveillance with radar devices [BSDH09]. Non-cooperative tracking comes with further related problems: There may be multiple objects so that the origins of measurements are unknown and clutter measurements that do not come from any object may occur. Non-cooperative tracking under such conditions is commonly known as *multiple target tracking* [BSWT11].

In general, each tracking system has to face the following requirements (see also [YSZ$^+$11]): First, it must be possible to track objects precisely

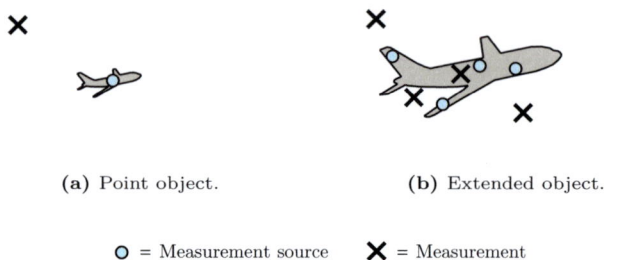

(a) Point object.   (b) Extended object.

○ = Measurement source   ✕ = Measurement

Figure 1.1: In contrast to a point object (a), an extended object (b) results in several measurements, where each measurement is a noisy observation of a measurement source on the object.

even under poor conditions, i.e., noise corrupted sensor data. Second, the tracking algorithm must be capable of following changes in the appearance of objects, e.g., shape changes, but also changes of the environment, e.g., landmarks may disappear. Third, the tracking algorithm should be efficient, e.g., real-time capable.

1.1 Considered Problem

Classical (non-cooperative) tracking methods are based on the assumption that the sensor noise is much larger than the spatial extent of the object (see Fig. 1.1) and hence, model it as a single point. However, due to the increasing resolution capability of modern sensor devices, the spatial extent of an object often cannot be neglected anymore and has to be incorporated into the tracking process.

Definition 1.1 (Extended Object). An *extended object* is characterized by a geometrically structured set of measurement sources that have a common dynamic behavior.

An *extended object* gives rise to several measurements per time instant. Each measurement is a noisy observation of a measurement source on the object, where the location of the measurement source is unknown (see Fig. 1.1). The measurement sources may vary over time and their locations depend on the object's shape but also on more complex properties such as the surface or the sensor's field of view.

Extended object tracking is a rather new problem that comes with a variety of challenges for a tracking system. This thesis focuses on the

(non-cooperative) tracking of a single extended object

based on noisy Cartesian point measurements.

Remark 1.1. The tracking of multiple extended objects and the incorporation of clutter measurements are beyond the scope of this thesis.

Remark 1.2. We restrict ourselves to Cartesian point measurements because most relevant sensors can be captured with point measurements, e.g., angle-distance measurements can be converted to point measurements.

Remark 1.3. Ignoring the spatial extent of an extended object, i.e., treating it as a point object, may cause overconfident and poor tracking results.

1.2 Basic Approach

In this thesis, the basic approach pursued is to approximate the extended object with a geometric shape such as an ellipse [GS05, KS05], [18]. As the parameters of the shape are unknown and may vary over time (e.g., the object may rotate), the tracking problem consists of the

simultaneous estimation of the kinematic and shape parameters

of the object.[1]

As usual in tracking applications, a recursive Bayesian state estimator [AMGC02, RAG04, BSWT11] is desired, which maintains a posterior probability density function for both the shape and the kinematic parameters. A Bayesian state estimator can be seen as the complete solution to the estimation problem as it encodes all available information in the posterior density. In particular, this information is required for further information processing problems such as gating, data association, or planning.

---

[1]In this thesis, the shape of an object is understood to be invariant to translations but not to rotations and scalings.

Of course, a suitable geometric shape for approximating the object highly depends on the particular scenario. In general, a more precise shape approximation is expected to increase the overall tracking performance due to the inherent connection between the shape and kinematic parameters. However, in case of high measurement noise, few available measurements, and a heavily maneuvering object, it may not be possible to extract all shape information. In this case, it is more suitable (or even necessary) to use a basic shape with few degrees of freedom in order to avoid track losses. In this manner, a coarsening of the object's shape is performed that neglects finer details that are difficult to estimate or even not required.

A suitable (rather informal) classification of relevant shapes can be performed according to the

- *dimension* of the shape, which the measurement sources are located on and

- the *description complexity* of the shape.

For instance, in two-dimensional space, the shape can be (see Fig. 1.2)

- zero-dimensional, i.e., a point object,

- one-dimensional, i.e., a curve such as an ellipse or a line, or

- two-dimensional, i.e., a closed curve with its interior (also called region).

The description complexity of the shape is classified into four different types (see Fig. 1.3):

- *Type 1*: Point
  The extended object is in fact a single point object.

- *Type 2*: Basic shape
  The shape of the extended object is a basic shape such as an ellipse or a line segment.

- *Type 3*: Connected shape
  The shape is connected, i.e., two points can be connected by a path.

(a) A closed curve, i.e., an ellipse (without interior).

(b) A region, i.e., an ellipse with interior.

Figure 1.2: One-dimensional (a) and two-dimensional (b) shape in two-dimensional space.

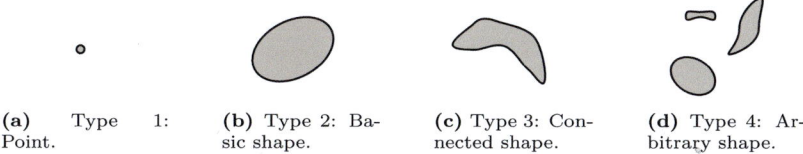

(a) Type 1: Point.

(b) Type 2: Basic shape.

(c) Type 3: Connected shape.

(d) Type 4: Arbitrary shape.

Figure 1.3: Description complexity of shapes.

- *Type 4*: General shapes
  There is no restriction on the shape, e.g., it may be composed of several basic shapes.

In the following, an N-dimensional shape in N-dimensional space is called a *region shape*. A one-dimensional shape (in N-dimensional space) is called a *curve*. In two-dimensional space, a region shape is given by a closed curve plus its interior. Please note that the distinction between region shapes and curves is essential in this thesis.

Remark 1.4. In the remainder of this thesis, we mainly restrict our discussions to the two-dimensional case. Nevertheless, generalizations to higher dimensions are usually straight-forward (if not stated otherwise).

Delimitation   *Contour tracking* [YSZ$^+$11, YJS06] refers to a class of tracking methods in computer vision, where objects are to be tracked in RGB image sequences. For example, the popular active contours [KWT88, JBU04, BI98] minimize an energy functional in order to find the object contour. Vision-based contour tracking algorithms have a large amount of data available, e.g., an RGB image, that covers the entire object. Hence,

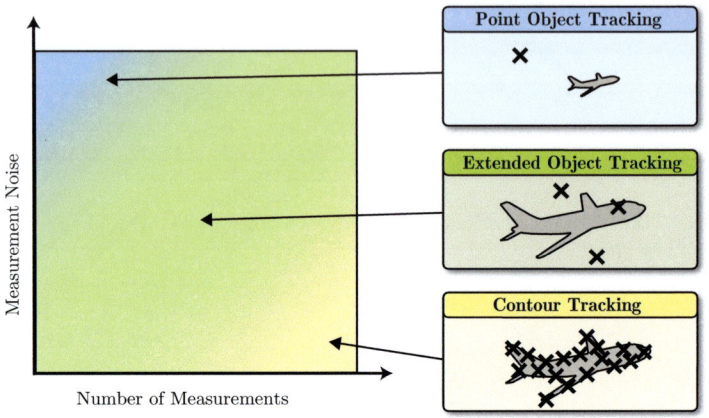

Figure 1.4: Taxonomy of shape tracking problems.

the challenge is rather *how* to obtain the measurements, i.e., the question is which image pixels belong to the object.

In a typical (single) extended object tracking scenario, the measurements are directly available and they are known to originate from the object (apart from clutter measurements). However, the available measurements at a specific time instant do not carry enough information for determining the contour of the object, i.e., there are too few or too noisy measurements given. The shape of the object can only be estimated if several time instants are considered under incorporation of the temporal evolution of the object. Contour tracking algorithms are in general not suitable for this kind of measurements as they are tailored to a large amount of RGB data with rather low noise that covers the entire object. Later in Chapter 6, it is shown experimentally that contour tracking algorithms may give poor results when applied to an extended object tracking problem.

According to the above discussion, *extended object tracking* can be roughly distinguished from classical *contour tracking* by means of the number of available measurements per frame and the magnitude of the measurement noise as depicted in Fig. 1.4.

1.3 Application Areas

Extended object tracking is a fundamental theoretical problem. Many sensor data fusion problems in robotics, surveillance, and computer vision can be cast as an extended object tracking problem. In general, each estimation problem involving noisy data sets arising from a structured set of unknown sources can be seen as an extended object tracking problem. Typically, extended objects occur in tracking applications when sensors such as radar devices, laser rangefinders, or optical devices are involved. For example, in [OGL11], the tracking of people with a laser rangefinder is treated, and in [GSDB07], vehicles are modeled as extended objects for the sake of collision detection. The works [JBC10, SCG09, JBC08] consider the tracking of contaminated clouds based on special sensors that measure the material concentration in the atmosphere. Finally, the processing of point cloud measurements obtained from depth sensors such as the Microsoft® Kinect™ can be interpreted as an extended object tracking problem [22].

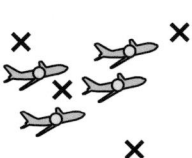

Figure 1.5: A group of point objects can be interpreted as a single extended object.

Extended object tracking methods can be employed for tracking a collectively moving group of point objects due to the strong relationship between individual group members (see Fig. 1.5). In this sense, a *group target* [WD04] is an extended object that consists of a finite set of measurement sources, i.e., the group members. Group object tracking applications can be widely found in robotics and surveillance. For instance, persons and person groups are modeled as clusters in [LAB09, CSG12] and, in [FK08, PD11], ground moving objects such as convoys are tracked with the help of a Ground Moving Target Indication (GMTI) sensor.

When the extended object is modeled as a curve, there is a strong relationship to curve fitting, which often occurs in computer vision, where curves shall be fitted to features extracted from image data [Por90, Zha97, FB08, ARR99]. For example, in [GML+08, GBL+11], a humanoid robot localizes spherical objects based on ellipse fitting methods. In addition to image data, fitting methods are vital for the processing of data supplied by a laser or radar device. For instance, mobile robots need circle fitting algorithms for outdoor localization using circular-shaped landmarks measured by a laser rangefinder [NVMBS08, ZXA06]. In [AOMMB07, COY10], ellipse fits serve for people tracking using also a laser rangefinder. The fitting of polynomial curves in a tracking context is treated in [LOG11], where road lanes are extracted from radar measurements.

1.4 Challenges

Single *extended* object tracking is a non-standard tracking problem that is significantly more difficult than single *point* object tracking. The two main challenges are the followings:

1. *High-dimensional, nonlinear estimation problem*
   A stochastic estimator for the shape of an extended object is based on a hierarchical nonlinear probabilistic model that first specifies the measurement sources and then the measurements. Even for simple shapes, the likelihood function for this model cannot be evaluated analytically, because the unknown measurement sources have to be marginalized out from the likelihood. Additionally, the state vector of an extended object is typically high-dimensional as it also contains shape parameters. All told, extended object tracking is a high-dimensional, nonlinear estimation problem for which standard approaches are bound to fail and elaborate approximation techniques are required.

2. *Lack of statistical information about measurement sources*
   It is often difficult or even impossible to obtain precise statistical information about possible locations of the measurement sources on the object as the true object shape and its properties are almost always totally unknown. A reasonable approach would be to assume that the measurement sources are uniformly distributed on the object's surface. However, then poor estimation results may be obtained if the measurement sources are in fact not uniformly distributed. For example, when tracking an airplane with a radar device, reflections from the turbines may be much more probable than elsewhere. A consequence of wrongly imposed statistical assumptions is that the estimation results can be significantly biased. Hence, a major challenge is to create a tracking algorithm that is robust to such systematic modeling errors.

Simultaneous Tracking and Shape Estimation of Extended Objects

Conic Fitting

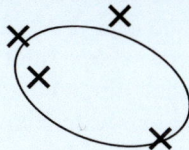

- Formulation as measurement equation with multiplicative noise
- Gaussian filter based on statistical linearization
- Suitable for large measurement noise

Random Hypersurface Model

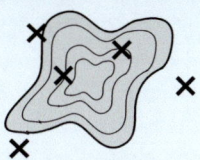

- Novel extent model for the shape interior
- Ellipses and star-convex shapes
- Gaussian filter for an efficient closed-form measurement update

Set-Theoretic Extent Model

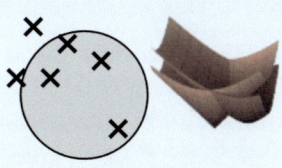

- No statistical assumptions on measurement sources
- Combined set-theoretic and stochastic estimator
- Robust to systematic errors

Experiments

- Ground moving object tracking
- Moving point detection using RGB and depth data
- Evaluation

Figure 1.6: Overview of this thesis.

1.5 Contributions and Outline of This Thesis

As extended object tracking is a nonlinear, high-dimensional estimation problem, existing approaches are usually based on particle filters and only basic shapes such as circles and line segments have been considered in literature (see the next section). For region shapes, closed-form formulas can only be found for the special case of ellipses with the help of random matrix theory [FFK11]. This thesis addresses this issue and demonstrates that extended object tracking is tractable; even detailed shape information can be estimated efficiently. For this purpose, Gaussian filters based on closed-form expressions are developed for a wide range of relevant shapes reaching from basic shapes to arbitrary free-form star-convex shapes. A new systematic approach for specifying region shapes called *Random Hypersurface Model* leverages the derivation of these filters.

The lack of statistical information about measurement sources has not been treated in literature at all. In this thesis, the negative effect of wrong assumptions in a stochastic estimator is illustrated and a set-theoretic extent model that does not make any statistical assumptions about the measurement sources is proposed. In this way, systematic errors in the measurement sources can be tackled with a mathematically sound formalism for incorporating unknown-but-bounded errors in Bayesian estimation theory.

The structure and main contributions of this thesis are outlined in Fig. 1.6. Each main topic is treated in a seperate chapter (Chapter 3 to Chapter 6). This and the last chapter form the introduction and conclusions of this thesis. Chapter 2 presents the mathematical framework and notation for simultaneously tracking and shape estimation of a single extended object with the help of Bayesian state estimators, where a strong focus lies on Gaussian filters. It serves as the basis for the remaining chapters, which are described in more detailed in the following.

- **Conic Fitting Using Statistical Linearization** – Chapter 3
  An important specific extended object tracking problem is obtained when the received point measurements originate from a closed curve, i.e., a circle or an ellipse (without the interior) as depicted in Fig. 1.2a. In this case, the problem can be interpreted as a "dynamic" fitting problem for which a recursive Bayesian state estimator is desired.

Although fitting problems are well-studied in literature, new requirements are imposed in the context of extended object tracking, e.g., the measurement noise is rather high and the conic evolves over time. In this thesis, we present novel recursive Gaussian filters for conic fitting that are tailored to the needs of extended object tracking. The basic approach is to reformulate the original implicit problem as a measurement equation that is corrupted by multiplicative noise. With the help of this explicit equation and statistical linearization, Gaussian filters that employ closed-form expressions for the measurement update are derived. In particular for large measurement noise, the novel fitting methods are able to significantly outperform state-of-the-art Gaussian-based approaches.

- **Random Hypersurface Model** – Chapter 4
  A main contribution of this thesis is a novel systematic approach for modeling a region shape, i.e., the interior of closed curve, which we call *Random Hypersurface Models (RHM)*. The basic idea is to reduce the modeling of a region shape to the modeling of a *curve* by means of scaling the shape boundary. In this vain, it is possible to derive explicit measurement equations for a wide class of relevant region shapes. In particular, we introduce an *RHM* for ellipses and free-form, i.e., arbitrary complex, star-convex shapes. Bayesian inference is performed with a Gaussian filter that allows for an efficient recursive measurement update based on *closed-form expressions*. These are the first Gaussian filters for region shapes, and most important, this is the first method at all that is able to estimate a star-convex shape approximation. By this means, it is possible to track an object whose shape is a priori totally unknown and estimated from scratch. This is a significant leap in extended object tracking as a detailed shape estimate is of high value for many applications, e.g., classification.

- **Set-Theoretic Extent Model** – Chapter 5
  In order to cope with the absence of statistical knowledge about the measurement sources, a set-theoretic extent model, which models the extent as an unknown-but-bounded error, is proposed, i.e., the only imposed assumption on a measurement source is that it lies on the object. For this set-theoretic extent model, we derive a combined set-theoretic and stochastic estimator that represents the

uncertainty of the state as a random set and uses (random) set intersection for the measurement update. Specifically, we develop novel outer-bounding techniques for circular-shaped extended objects. The set-theoretic extent model implicates that the object's extent cannot be estimated only with point measurements. However, it can be inferred from further information sources, e.g., the number of received measurements. Simulations demonstrate that the set-theoretic approach may yield more robust and precise estimation results than a pure stochastic approach in case the locations of the measurement sources are dominated by a systematic error.

- **Experiments: Moving Object Tracking using RGBD Data –** Chapter 6
  We present an experimental setup for evaluating extended object tracking methods. In this experiment, a moving object is to be tracked with the help of an RGBD camera observing the scene from a bird's eye view. For this purpose, moving points are detected in the RGB and depth image sequences. The extracted moving points originate from the surface of the moving object and serve as input for the extended object tracking algorithm. With the help of this experiment, we provide an exhaustive evaluation of *RHMs* for star-convex shapes. We illustrate the need of an extended object tracking method for this scenario and compare *RHMs* with active contour models, which are a state-of-the-art approach for vision-based contour tracking.

1.6 Related Work

In this section, a brief outline of related extended object tracking methods is given. More detailed discussions of relevant state-of-the-art methods with respect to this thesis are given in the corresponding chapters. In general, related methods can be roughly subdivided into

- curve fitting methods,
- approaches that represent the object as a finite set of measurement sources, and
- approaches for region shapes, i.e., closed curves with an interior.

Curve Fitting   In case the measurement sources lie on a curve, the problem can be seen as a curve fitting problem [Che10]. Fitting curves such as circles, ellipses and line segments to noisy data is a traditional problem that is still an active research area in computer vision and robotics. A huge amount of solution methodologies (statistical and non-statistical) for this problem have been developed. Curve fitting usually denotes a static problem, i.e., the best fitting curve for several measurements is to be found. In this thesis, we have to deal with a dynamic problem, i.e., a tracking problem, for which we aim at a recursive Bayesian state estimator. A Bayesian approach to curve fitting can be found in [WK01] and in [Por90], Kalman filtering techniques are used for ellipse fitting. A more detailed discussion of these curve fitting methods is given in Section 3.2.

Explicit Extent Models   The classical approach to extended object and group tracking is to represent the object explicitly with a finite set of measurement sources that share a bulk component in common, e.g., the velocity [BC91, IG03, VIG05, VIG04, GSDB07]. The goal is to estimate both the locations of the measurement sources and the bulk component. This explicit object model is suitable when reasonable models for the measurement sources are available and each single measurement source can be resolved by the sensor. However, there are several situations in which explicit models are not suitable and an implicit model is advantageous:

- Explicit models for the point features, e.g., for the motion and detection probability, may not be available and not be justified.

- In case of an extended object, where the measurement sources originate from a continuous set, finite sets of point features are a poor model of the real object.

- Resolution conflicts may render it impossible to resolve single point features.

- With an increasing number of point features, the problem may become intractable as the number of association hypothesis grows exponentially with the number of measurement sources.

Region Shapes   The plain Bayesian approach for estimating region shapes is to model the shape as a so-called spatial probability distribution [GS05,

GGMS05], i.e., each measurement source is a random draw from a specific probability distribution whose probability mass is concentrated on the object.

In literature, particle filters based on spatial distributions have been used for estimating region shapes such as circles [WK01, PMGA12, PGMA12] and Gaussian mixtures with known parameters [GS05]. Of course, spatial distributions can been also used for estimating curves, e.g., circles [WK01, PMGA11] and stick targets [GS05, GGMS05, BDT+06]. Spatial distributions have been integrated into the *Probability Hypothesis Density (PHD)* filter framework for tracking multiple extended objects [Mah09, GLO10, SC10, GO12].

In [KS05, Koc08], a spatial distribution is used to model elliptic shapes. For this purpose, the uncertainty about the ellipse is specified by means of a random symmetric positive definite matrix. Each measurement source is assumed to be a random draw from a Gaussian distribution whose covariance matrix is specified by the random matrix. The original approach [KS05,Koc08] neglects the measurement noise but an extension that incorporates the measurement noise was proposed in [FF08,FF09,FFK11], which, however, comes with a further loss of optimality. Although the random matrix approach assumes measurement sources to be drawn from a Gaussian distribution, uniformly distributed measurement sources on an ellipse can be captured through moment matching [FF09, FFK11]. Multiple extended objects have been considered within the *Probabilistic Multiple-Hypothesis Tracker (PMHT)* framework [CGW+09] in [WK10, WK12] and PHD filters [GO12]. Very recent further developments can be found in [Org12, LRL12a, LRL12b]. Besides of the methods proposed in this thesis, the random matrix approach is currently the only approach that gives closed-form expressions for the measurement and time update. A comparison of *RHMs* for ellipses and the random matrix approach can be found in [6] and Section 4.2.4.

CHAPTER 2

A Bayesian Framework for Extended Object Tracking

This chapter is about a Bayesian framework for tracking a *single extended object* based on sequentially arriving *noisy point measurements*.

First, the required probabilistic models for an extended object are introduced. Essentially, we consider independently generated noisy point measurements originating from unknown measurement sources on an extended object. As the shape of the object is unknown and may vary over time, the state vector is enhanced with parameters that determine the shape.

Second, a conceptual recursive Bayesian estimator for the state, i.e., the kinematic and shape parameters, is derived based on the probabilistic model. A Bayesian state estimator computes a posterior probability density function for the state vector based on all available information, i.e., the measurements [AMGC02]. As the estimation problem is in general nonlinear, high-dimensional, and already the likelihood function cannot be evaluated analytically, standard (linear) estimation techniques cannot be applied.

After discussing the difficulties with a naïve particle filter implementation for extended objects, we recall the well-known concept of statistical linearization, which will be used in the remainder of this thesis for deriving specific extended object trackers. The main benefits of statistical linearization in this context are that the likelihood function does not have to be

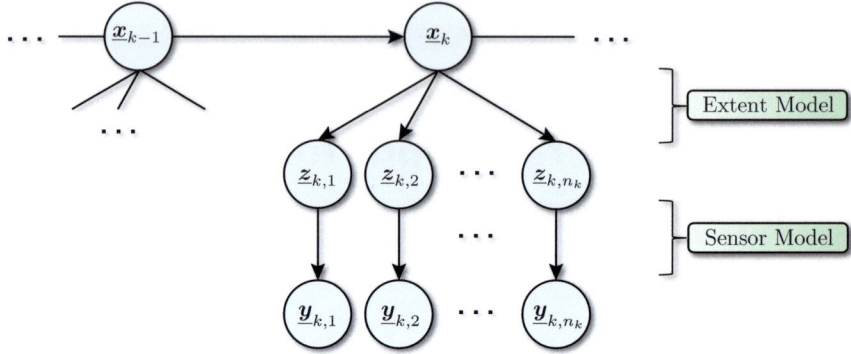

Figure 2.1: Graphical model visualizing the dependencies of the involved quantities for extended object tracking, i.e., the state vector $\underline{x}_k$, the measurements $\underline{y}_{k,1}, \underline{y}_{k,2}, \dots, \underline{y}_{k,n_k}$, and the measurement sources $\underline{z}_{k,1}, \underline{z}_{k,2}, \dots, \underline{z}_{k,n_k}$.

evaluated explicitly and closed-form expressions can be derived for a wide class of measurement equations such as polynomial equations.

All extended object tracking methods proposed in the subsequent chapters are based on this chapter. However, it is important to note that the combined set-theoretic and stochastic estimator from Chapter 5 is an extension of the concept of a Bayesian state estimator.

Remark 2.1. The Bayesian formulation of the extended object tracking problem is frequently used in literature, see for example [GS05, GGMS05, BDT$^+$06, Koc08, PMGA11] and also [3], [9, 14, 18, 19]. Here, we follow the notation of [3], [9, 14, 18, 19] that focuses on a single extended object without false measurements.

2.1   Modeling Extended Objects

The extended object tracking problem is formulated as a discrete-time stochastic dynamic system (see for example [GS05, GGMS05, Koc08] and [9, 14, 18, 19]). For this purpose, it is necessary to specify

- the object state,

- a measurement model, which relates the state to the measurements, and

- a system model for the temporal evolution of the state.

A graphical model of the involved quantities is depicted in Fig. 2.1. The state vector is given by $\underline{\boldsymbol{x}}_k$, where k denotes the time step. It contains parameters that specify both the kinematics and the shape of the object. At each time step k, n_k independent measurements $\mathcal{Y}_k = \{\underline{\boldsymbol{y}}_{k,l}\}_{l=1}^{n_k}$ are available, where each measurement $\underline{\boldsymbol{y}}_{k,l}$ originates from a particular measurement source $\underline{\boldsymbol{z}}_{k,l}$ on the object, see also Fig. 2.2. Hence, the measurement model involves two components, i.e., the *extent model*, which specifies a measurement source for given object state, and the *sensor model*, which determines the measurement based on the measurement source.

As visualized in the graphical model Fig. 2.1, single measurements are mutually independent for given object state, i.e., a single measurement only depends on the measurement source and the measurement source only depends on the state vector. A measurement model for a single measurement is in particular required in case only a single measurement per time instant is a available.

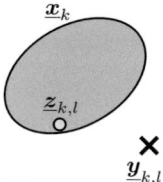

Figure 2.2: Measurement model for extended objects: Independent generation of measurements.

Further details about the involved quantities and the relations between them are described in the following.

Remark 2.2. We denote a particular realization of $\underline{\boldsymbol{y}}_{k,l}$ with $\hat{\underline{\boldsymbol{y}}}_{k,l}$ and define the abbreviation $\hat{\mathcal{Y}}_k = \{\hat{\underline{\boldsymbol{y}}}_{k,l}\}_{l=1}^{n_k}$.

2.1.1 State Vector

The object state at time k is represented with a random vector $\underline{\boldsymbol{x}}_k = \left[\underline{\boldsymbol{m}}_k^T, (\underline{\boldsymbol{x}}_k^*)^T, \underline{\boldsymbol{p}}_k^T\right]^T$ that encompasses

- the N-dimensional center of the object $\underline{m}_k$,

- variables $\underline{x}_k^*$ for the kinematic parameters such as the velocity, and

- the N^p-dimensional parameter vector $\underline{p}_k$ that specifies the shape.

The shape parameter $\underline{p}_k$ is a parametric representation of a geometrically structured set in $\mathbb{R}^N$ that is denoted as $\mathcal{S}(\underline{p}_k) \subset \mathbb{R}^N$. Note that the center of the object is not part of the shape parameters, i.e., in this thesis the shape is invariant to translation but *not* to rotation and scaling.

Example 2.1 (Circle). A circular disc can be represented with $\underline{p}_k = r_k$, where r_k is the radius. The shape is given by $\mathcal{S}(\underline{p}_k) = \{\underline{z} \mid \underline{z} \in \mathbb{R}^2$ and $||\underline{z}||_2 \leq r_k\}$.

The entire set of measurement sources associated to an extended object is denoted as

$$\mathcal{M}(\underline{x}_k) = \mathcal{M}(\underline{p}_k, \underline{m}_k) := \mathcal{S}(\underline{p}_k) + \underline{m}_k \ , \qquad (2.1)$$

where the center $\underline{m}_k$ is added to each element of $\mathcal{S}(\underline{p}_k)$.

2.1.2   Measurement Model

As we assume that the single measurements are mutually independent, a measurement model for a single measurement is sufficient. Hence, the measurement model relates the state vector $\underline{x}_k$ to a measurement $\underline{y}_{k,l}$, where it also incorporates that the measurement stems from an unknown measurement source $\underline{z}_{k,l}$ on the extended object. It is composed of two parts, the *extent model* and the *sensor model* (see Fig. 2.2).

Extent Model   For a given extended object $\mathcal{M}(\underline{p}_k, \underline{m}_k)$, the extent model specifies a *measurement source* $\underline{z}_{k,l} \in \mathcal{M}(\underline{p}_k, \underline{m}_k)$ (see Fig. 2.2). Often little information about possible locations of a measurement source is available. At least it is known that a measurement source lies on the extended object.

A reasonable extent model is a spatial distribution [GS05, GGMS05], i.e., each measurement source $\underline{z}_{k,l}$ is a random draw of the probability distribution

$$f(\underline{z}_{k,l} \mid \underline{x}_k) = f(\underline{z}_{k,l} \mid \underline{p}_k, \underline{m}_k) \ .$$

An obvious choice for a spatial distribution is the uniform distribution, which says that each measurement source is equally probable, i.e.,

$$f(\underline{z}_{k,l} \mid \underline{p}_k, \underline{m}_k) = \frac{1}{|\mathcal{S}(\underline{p}_k)|} \ ,$$

where $|\mathcal{S}(\underline{p}_k)|$ denotes the area of the shape/length of the curve $\mathcal{S}(\underline{p}_k)$. In this thesis, two further extent models are introduced, i.e.,

- the *Random Hypersurface Model*, which assumes that each measurement source lies on a scaled version of the shape boundary (see Chapter 4), and the

- *set-theoretic extent model*, which does not make any assumptions on a measurement source except that it lies on the object shape (see Chapter 5).

It is essential to note that the locations of the measurement sources are not desired. They are just a required (unknown) quantity on the route to the measurement.

Sensor Model   Suppose we are given a measurement source $\underline{z}_{k,l}$, then the sensor model says how to get the measurement $\underline{y}_{k,l}$. As we focus on Cartesian point measurements corrupted with additive Gaussian noise, the measurement equation is

$$\underline{y}_{k,l} = \underline{z}_{k,l} + \underline{v}_{k,l} \ , \tag{2.2}$$

where $\underline{v}_{k,l}$ is zero-mean white Gaussian noise with covariance $\Sigma^v_{k,l}$. Based on (2.2), the likelihood function

$$f(\underline{\hat{y}}_{k,l} \mid \underline{z}_{k,l}) = \mathcal{N}(\underline{\hat{y}}_{k,l} - \underline{z}_{k,l}, \Sigma^v_{k,l})$$

is obtained for a particular measurement $\underline{\hat{y}}_{k,l}$.

Note that almost all relevant sensors can be captured within this model. For example, angle-distance measurements can be converted to Cartesian point measurements as described in [Frä07].

Extensions of the Model   The model introduced above can be seen as a minimal model for tracking a single extended object. Of course, it can be further extended. For example, it may be suitable to consider the number of received measurements n_k as a random variable, where the number may depend on the size of the extended object.

2.1.3   System Model

The temporal evolution of the object state $\underline{x}_k$ is modeled as a Markov process. For an extended object, it also specifies how the shape parameters evolve over time. In this thesis, no restrictions on the system model are imposed. Typically, it is specified by a system equation such as

$$\underline{x}_k = \underline{a}_{k-1}(\underline{x}_{k-1}, \underline{u}_{k-1}, \underline{w}_{k-1}) \ , \tag{2.3}$$

where $\underline{a}_{k-1}(\cdot, \cdot, \cdot)$ denotes the system function, $\underline{u}_{k-1}$ is the system input, and $\underline{w}_{k-1}$ is the (mutually independent) system noise. Based on the system equation (2.3), a conditional density function $f(\underline{x}_k|\underline{x}_{k-1})$ can be derived [Sim06, RAG04].

Although a general nonlinear system model such as (2.3) may be used, we focus in this thesis on two simple linear motion models, which are the *random walk* model and *nearly constant velocity* model.

Random Walk Model   A very simple motion model is the so-called random walk model, which just perturbs the state vector with an additive zero-mean Gaussian noise term. This model is reasonable for the center of the object but also for the shape parameters. By this means it is possible to capture small shape changes during two time instants. Here, we consider the random walk model given by

$$\underline{x}_k = \underline{x}_{k-1} + \underline{w}_{k-1} + \underline{\hat{u}}_{k-1} \ , \tag{2.4}$$

where $\underline{w}_{k-1}$ is zero-mean white Gaussian noise and $\underline{\hat{u}}_{k-1}$ is a deterministic input vector.

Constant Velocity Model for the Center   A constant velocity model [BSKL02] for the center assumes that the object moves approximately along a straight line with a constant velocity. However, the shape parameters are still modeled as a random walk.

In the following, let the state vector be $\underline{x}_k = \left[\underline{m}_k^T, (\underline{m}_k^v)^T, \underline{p}_k^T \right]^T$, where $\underline{m}_k^v$ is the velocity vector. In two-dimensional space, the nearly constant velocity model [BSKL02] for the center of the object is given by

$$\underline{x}_k = \mathbf{A}_{k-1}\underline{x}_{k-1} + \begin{bmatrix} \mathbf{B}_{k-1}^{cv}\underline{w}_{k-1}^{cv} \\ \underline{w}_{k-1}^{p} \end{bmatrix} , \tag{2.5}$$

where

- $\mathbf{A}_{k-1} = \operatorname{diag}(\mathbf{A}_{k-1}^{cv}, \mathbf{I}_{N^p})$ with $\mathbf{A}_{k-1}^{cv} = \begin{bmatrix} 1 & 0 & T & 0 \\ 0 & 1 & 0 & T \\ 0 & 0 & 1 & 0 \\ 0 & 0 & 0 & 1 \end{bmatrix}$ and $\mathbf{I}_{N^p}$ is the identity matrix with dimension N^p,

- $\mathbf{B}_{k-1}^{cv} = \begin{bmatrix} \frac{1}{2}T^2 & 0 \\ 0 & \frac{1}{2}T^2 \\ T & 0 \\ 0 & T \end{bmatrix}$ is a matrix,

- $\underline{w}_{k-1}^{cv}$ is zero-mean Gaussian noise for the constant velocity model, whose covariance matrix is a design parameter, and

- $\underline{w}_{k-1}^{p}$ is zero-mean Gaussian noise modeling small changes in the shape parameters.

Note that there is a lot of room left for future research concerning more elaborate motion models for extended objects.

2.2 Formal Bayesian State Estimator for Extended Objects

Based on the probabilistic measurement and system model, a (formal) recursive Bayesian state estimator for the state $\underline{x}_k$ can be derived (see also [GS05, GGMS05, Koc08], [9, 14, 18, 19]). As an extended object may give rise to multiple measurements per time instant, the standard notation for a Bayesian state estimator will be slightly adapted, i.e., the posterior probability density for the state vector $\underline{x}_k$ having incorporated the measurements $\hat{\underline{y}}_{k,1}, \ldots, \hat{\underline{y}}_{k,l}$ (and all measurements from previous time steps) with $f_l(\underline{x}_k)$, i.e.,

$$f_l(\underline{x}_k) := f\left(\underline{x}_k \mid \hat{\mathcal{Y}}_1, \ldots, \hat{\mathcal{Y}}_{k-1}, \hat{\underline{y}}_{k,1}, \ldots, \hat{\underline{y}}_{k,l}\right) .$$

A recursive Bayesian state estimator consists of a time and measurement update step. Although multiple measurements are received per time step, they can be incorporated recursively as they are independent (see Section 2.1.2). In this vein, a time update may be followed by a sequence of measurement updates.

2.2.1   Time Update

Suppose we are given the probability density $f_{n_{k-1}}(\underline{x}_{k-1})$ for the time step $k-1$ after incorporating all n_{k-1} measurements, then the time update results in the probability density $f_0(\underline{x}_k)$ for the predicted state for time instant k. The prediction $f_0(\underline{x}_k)$ results from the well-known Chapman-Kolmogorov equation

$$f_0(\underline{x}_k) = \int f(\underline{x}_k \mid \underline{x}_{k-1}) \cdot f_{n_{k-1}}(\underline{x}_{k-1}) \, \mathrm{d}\underline{x}_{k-1} . \qquad (2.6)$$

2.2.2   Measurement Update

In the measurement update step, the prediction $f_0(\underline{x}_k)$ is updated with the measurements $\hat{\mathcal{Y}}_k = \{\hat{\underline{y}}_{k,l}\}_{l=1}^{n_k}$ based on Bayes' rule

$$f_{n_k}(\underline{x}_k) = \alpha_k \cdot f\left(\hat{\mathcal{Y}}_k \mid \underline{x}_k\right) \cdot f_0(\underline{x}_{k-1}) ,$$

where $f\left(\hat{\mathcal{Y}}_k \mid \underline{x}_k\right)$ is the likelihood function derived from the measurement model and α_k is a normalization factor. As the measurements are mutually independent (for a given state), they can be incorporated recursively

$$f_l(\underline{x}_k) = \alpha_{k,l} \cdot f(\hat{\underline{y}}_{k,l} \mid \underline{x}_k) \cdot f_{l-1}(\underline{x}_k) ,$$

where $f(\underline{\hat{y}}_{k,l}|\underline{x}_k)$ is the single measurement likelihood function and $\alpha_{k,l}$ is again a normalization factor. For a spatial distribution model, the single measurement likelihood function is

$$f(\underline{\hat{y}}_{k,l}|\underline{x}_k) = \int f(\underline{\hat{y}}_{k,l}|\underline{z}_{k,l}) \cdot f(\underline{z}_{k,l}|\underline{x}_k)\, \mathrm{d}\underline{z}_{k,l} \ , \qquad (2.7)$$

which results from marginalizing out the unknown measurement source. Note that the measurement does only depend on the measurement source, i.e., $f(\underline{\hat{y}}_{k,l}|\underline{z}_{k,l}, \underline{x}_k) = f(\underline{\hat{y}}_{k,l}|\underline{z}_{k,l})$.

2.2.3 Discussion

Typically, the measurement model for extended objects is a highly non-linear, hierarchical probability model. As a consequence, the probability density $f_l(\underline{x}_k)$ for the state cannot be calculated recursively in closed-form. Maintaining an approximation of $f_l(\underline{x}_k)$ is a challenging problem as

- the state vector is usually high-dimensional, and

- the likelihood function (2.7) cannot be evaluated analytically even for simple basic shapes.

In the remaining two sections of this chapter, we discuss two approaches for dealing with this nonlinear estimation problem.

2.3 A Naïve Particle Filter for Extended Object Tracking

Particle filters employ sequential Monte Carlo techniques for maintaining a point mass representation of the posterior density in a nonlinear Bayesian estimator. For a detailed overview of recent methods, we refer to [AMGC02, RAG04].

As every nonlinear estimation technique, particle filter approaches come with particular advantages and disadvantaged. For the considered problem, the naïve application of a particle filter is rather challenging. In order to show the difficulties, we define a naïve particle filter called *Naïve-PF*, which will serve as a benchmark filter in the remainder of this thesis.

Remark 2.3. The *Naïve-PF* for extended objects is characterized as follows:

- The likelihood function (2.7) for the measurement update is approximated by performing a random sampling of the curve/region.

$$f\left(\underline{\hat{y}}_{k,l}|\underline{x}_k\right) \approx \sum_{i=1}^{N_z} f\left(\underline{\hat{y}}_{k,l}|\underline{z}_{k,l}^{(i)}\right) \cdot f\left(\underline{z}_{k,l}^{(i)}|\underline{x}_k\right) \ ,$$

  where $\underline{z}_{k,l}^{(i)}$ with $i = 1, \ldots, N_z$ are random samples from the extended object $\mathcal{M}(\underline{p}_k, \underline{m}_k)$.

- After each measurement update, a simple resampling is performed by sampling from a Gaussian whose mean and covariance are obtained from matching the moments of the posterior samples. This resampling step is in particular required for estimating stationary extended objects.

The above described *Naïve-PF* can be seen as a Gaussian particle filter [KD03] tailored to extended objects. The difficulties with the *Naïve-PF* are exemplified with an elliptic shape. For an ellipse, the state vector may be seven dimensional (five parameters for the ellipse plus two for the velocity vector). Using ten samples per dimension results in an overall number of 10^7 samples for the probability density of the state vector. In order to approximate the likelihood function (2.7), it is reasonable to choose again ten samples per dimension, i.e., overall 10^2 samples for one evaluation of the likelihood. In summary, 10^9 particles have to be considered just for multiplying the likelihood function with the weights of the prior samples. Additionally, one has to deal with typical particle filter problems such as particle degeneration.

2.4   Gaussian Filter Using Statistical Linearization

In this thesis, we advocate Gaussian filters based on statistical linearization [Sim06, JU04, HH08, BHH10, LABS06]. A Gaussian filter approximates the exact probability distribution of the state with a Gaussian distribution and statistical linearization yields a linear approximation of a nonlinear (measurement or system) function by incorporating statistical information. Based on the linear approximation, the Kalman filter formulas can be used for the measurement and time update.

The main benefits of statistical linearization are outlined in the following:

- It is not necessary to directly evaluate the likelihood (2.7).

- Even for high-dimensional states, fast and accurate approximate statistical linearization methods are available. For example, *Linear Regression Kalman Filters* such as the *UKF* [JU04] and [HH08] perform statistical linearization with the help of deterministic sampling.

- For several relevant special cases such as polynomials, statistical linearization can be performed analytically in closed-form by means of moment matching.

- Due to the Gaussian distribution, there is no particle degeneration as in particle filters.

- Gaussian filters can be directly embedded into other well-established frameworks that are based on Gaussians such as multiple model approaches or multiple target trackers (see [BSWT11]).

In the following, we recall the concept of statistical linearization and then apply it to the measurement and time update step.

2.4.1 Statistical Linearization

The statistical linearization [LBDS04] of a nonlinear function $\underline{G} : \mathbb{R}^N \rightarrow \mathbb{R}^M$ with

$$\underline{G}(\underline{x}) = \underline{y}$$

incorporates that $\underline{x} \sim \mathcal{N}(\underline{x} - \underline{\mu}^x, \Sigma^{xx})$ is a random vector. For this purpose, the joint density of $\underline{y}$ and $\underline{x}$ is approximated with a Gaussian density, i.e., the mean $\underline{\mu}^y$, cross-covariance Σ^{xy}, and covariance Σ^{yy} are determined. Based on the approximation of the joint density, the linear approximation of $\underline{G}(\cdot)$ turns out to be

$$\underline{y} \approx \mathbf{A}_G \cdot \underline{x} + \underline{b}_G \ ,$$

where

$$\mathbf{A}_G = (\Sigma^{xy})^T \Sigma^{xx} \, , \tag{2.8}$$
$$\underline{b}_G = \underline{\mu}^y - \mathbf{A}_G \underline{\mu}^x \, . \tag{2.9}$$

In this thesis, the approximation of the joint density is performed by means of moment matching, see also [Sim06,JU04,HH08,BHH10,LABS06]. In this manner, a so-called *Linear Minimum Mean Squared Estimator (LMMSE)* is obtained. Usually, only an approximation of the moments is calculated. For instance, [JU04, HH08] propagate deterministically chosen samples through the measurement function. Nevertheless, for some special cases such as polynomial functions analytic expressions for the exact moments are available, see [BHH10, LABS06]. In the following, the efficient general moment calculation procedure for a polynomial function discussed in [17] is described.

Moments of a Polynomial Transformed Gaussian Random Vector

Let

$$\underline{G}(\underline{x}) = \begin{bmatrix} G_1(\underline{x}) \\ \vdots \\ G_M(\underline{x}) \end{bmatrix}$$

be an M-dimensional *polynomial* function with component functions $G_l(\underline{x})$ written as sum of products

$$G_l(\underline{x}) = \sum_j a_{j,l} \prod_{i=1}^{N} x_i^{s_{i,j,l}} \text{ for } 1 \leq l \leq M \, ,$$

where x_i is the i-th component of $\underline{x}$, $s_{i,j,l}$ is the exponent of x_i in the j-th sum of the l-th component function, and $a_{j,l}$ is the coefficient of the j-th sum in the l-th component function.

The mean $\underline{\mu}^y$ of $G(\underline{x})$ can be written as

$$\underline{\mu}^y = \begin{bmatrix} \mu_1^y \\ \vdots \\ \mu_M^y \end{bmatrix} = \mathrm{E}\{G(\underline{x})\} = \begin{bmatrix} \mathrm{E}\{G_1(\underline{x})\} \\ \vdots \\ \mathrm{E}\{G_M(\underline{x})\} \end{bmatrix} =$$

$$\begin{bmatrix} \sum_j a_{j,1} \, \mathrm{E}\left\{ \prod_{i=1}^N x_i^{s_{i,j,1}} \right\} \\ \vdots \\ \sum_j a_{j,M} \, \mathrm{E}\left\{ \prod_{i=1}^N x_i^{s_{i,j,M}} \right\} \end{bmatrix} . \quad (2.10)$$

In the same manner, the covariance matrix $\Sigma^y = (\sigma_{l,m}^2)_{l,m=1,\dots,M}$ of $G(\underline{x})$ is composed of

$$\sigma_{l,m}^2 = \mathrm{E}\{G_l(\underline{x}) \cdot G_m(\underline{x})\} - \mu_l^y \mu_m^y .$$

Because the product $G_l(\underline{x}) \cdot G_m(\underline{x})$ can be expanded to a polynomial in the form $\sum_j a_{j,(l,m)} \prod_{i=1}^N x_i^{s_{i,j,(l,m)}}$, we obtain

$$\sigma_{l,m}^2 = \sum_j a_{j,(l,m)} \, \mathrm{E}\left\{ \prod_{i=1}^N x_i^{s_{i,j,(l,m)}} \right\} - \mu_l^y \mu_m^y . \quad (2.11)$$

In this vein, the computation of the overall mean (2.10) and covariance matrix (2.11) of $G(\underline{x})$ has been broken down to the expectation of products of *non-central dependent* Gaussian random variables. A formula for the expectation of products of (zero-mean) random variables is available in literature for a long time due to Isserlis [Iss18]. However, as the Isserlis formula is already for small exponentials intractable, we propose to employ a computationally more attractive alternative formula derived in [Kan08].

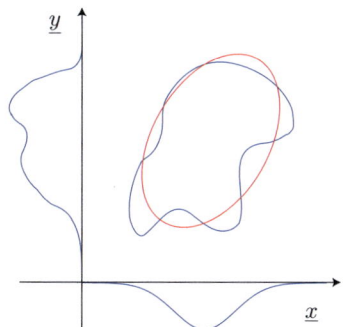

Figure 2.3: Statistical linearization: Marginal densities and joint density (blue) of $\underline{y}$ and state $\underline{x}$. Statistical linearization is performed by approximating the joint density with a Gaussian density (red).

Theorem 2.1. *For a Gaussian random vector* $\underline{x} = [\underline{x}_1, \ldots, \underline{x}_N]^T \sim \mathcal{N}(\underline{x} - \underline{\mu}^x, \Sigma^x)$ *and nonnegative integers* s_1 *to* s_N, *the following holds* [Kan08]:

$$
\mathrm{E}\left\{ \prod_{i=1}^{N} x_i^{s_i} \right\} = \sum_{\nu_1=0}^{s_1} \cdots \sum_{\nu_N=0}^{s_N} \sum_{r=0}^{\lfloor s/2 \rfloor} (-1)^{\sum_{i=1}^{N} \nu_i} \cdot \binom{s_1}{\nu_N} \cdots \binom{s_n}{\nu_N}
$$

$$
\cdot \frac{\left(\frac{\underline{h}^T \Sigma^x \underline{h}}{2} \right)^r \left(\underline{h}^T \underline{\mu}^x \right)^{s-2r}}{r!(s-2r)!} \quad , \quad (2.12)
$$

where $\underline{h} = \left[\frac{s_1}{2} - \nu_1, \ldots, \frac{s_N}{2} - \nu_N \right]^T$ *and* $s = s_1 + \ldots + s_N$.

All told, the first two moments of the polynomial transformation of a Gaussian random vector can be calculated analytically as follows:

1. Formulate the first two moments of $\underline{G}(\underline{x})$ in terms of the expectation of sums of products as in (2.10) and (2.11).

2. Determine the expectation of the sum of products using Theorem 2.1.

Step 2 is computationally expensive for higher order polynomials. However, the polynomial function (2.10) will only be quadratic in this thesis so that all moments can be calculated efficiently.

Remark 2.4. The above procedure does not simplify the resulting formulas. For quadratic equations, it is possible to obtain more compact expressions when exploiting well-known identities such as in [PP08, Bro11]. As these derivations are straightforward but tedious, we mostly refer to the general procedure described above.

2.4.2   Measurement Update

In order to employ statistical linearization for the measurement update, we assume that a measurement function $h_k(\underline{x}_k, \underline{v}_{k,l})$, which maps the state $\underline{x}_k$ and a noise term $\underline{v}_{k,l}$ to the measurement $\underline{y}_{k,l}$ is available, i.e.,

$$
\underline{y}_{k,l} = h_k(\underline{x}_k, \underline{v}_{k,l}) \ . \quad (2.13)
$$

Remark 2.5. In this work, the measurement function (2.13) will be derived with the help of the extent and sensor model (see Section 2.1). How *suitable* measurement functions for extended objects can be obtained is not obvious and will be discussed in the subsequent chapters.

The measurement update [Sim06, JU04, HH08, BHH10, LABS06] is performed in the following two substeps (see Fig. 2.3).

- *Substep 1: Statistical linearization*
  In the first substep, the statistical linearization of the measurement function is performed, i.e., for given $f_{l-1}(\underline{x}_k) = \mathcal{N}(\underline{x}_k - \underline{\mu}_{k,l-1}^x, \Sigma_{k,l-1}^x)$, the joint distribution

$$\begin{bmatrix} \underline{x}_k \\ \underline{y}_{k,l} \end{bmatrix} = \begin{bmatrix} \underline{x}_k \\ h_k(\underline{x}_k, \underline{v}_{k,l}) \end{bmatrix}$$

  is approximated with a Gaussian distribution with mean $\begin{bmatrix} \underline{\mu}_{k,l-1}^x \\ \underline{\mu}_{k,l}^y \end{bmatrix}$ and covariance matrix $\begin{bmatrix} \Sigma_{k,l-1}^x & \Sigma_{k,l}^{yx} \\ \Sigma_{k,l}^{xy} & \Sigma_{k,l}^y \end{bmatrix}$.

- *Substep 2: Update based on the Kalman filter formulas*
  The second substep consists of determining the updated estimate and covariance matrix by conditioning the Gaussian approximation of the joint density on the particular measurement $\hat{\underline{y}}_{k,l}$ using the Kalman filter equations

$$\begin{aligned} \underline{\mu}_{k,l}^x &= \underline{\mu}_{k,l-1}^x + \Sigma_{k,l}^{xy} \left(\Sigma_{k,l}^{yy} \right)^{-1} \left(\hat{\underline{y}}_{k,l} - \underline{\mu}_{k,l}^y \right) \\ \Sigma_{k,l}^x &= \Sigma_{k,l-1}^x - \Sigma_{k,l}^{xy} \left(\Sigma_{k,l}^{yy} \right)^{-1} \Sigma_{k,l}^{yx} . \end{aligned}$$

At this point, it is essential to realize that statistical linearization has to be applied with care as the approximation of the joint density with a Gaussian may be a rough approximation, i.e., the linearization error may be high. A naïve application of statistical linearization to extended object tracking problems is not promising due to strong nonlinearities in the measurement model. For example, later in Section 3.4.1, it is shown that naïve statistical linearization is unsuitable for circle fitting due to the poor approximation accuracy. Furthermore, it has been proven in [4] that it is impossible to estimate the parameters of an elliptic region based on a naïve statistical linearization.

2.4.3   Time Update

As we only consider linear motion models in this thesis, the time update can directly be performed with the help of the standard Kalman filter formulas [Sim06]. However, in general, a statistical linearizing of the system function (2.3) is possible in the same manner as for the measurement function.

2.5   Conclusions

This chapter introduced the basic models and estimation techniques required for the (single) extended object tracking methods developed in this thesis.

Since the shape of the extended object is unknown and part of the estimation problem, the state vector of an extended object contains shape parameters in addition to kinematic parameters. The measurement model incorporates that each measurement stems from an unknown measurement source on the object, where it is assumed that measurements are mutually independent. This independence assumption allows for inferring the shape when only a single measurement per time instant is available.

The temporal evolution of an extended object is specified by a system model for both the kinematic and shape parameters. In this thesis, we restrict ourselves to simple linear motion models, e.g., a nearly constant velocity model for the center and a random walk model for the shape parameters.

Already single extended object tracking based on the introduced probabilistic model is in general highly nonlinear and computationally intractable. We discussed two approximation techniques for a recursive Bayesian estimator for the object state, i.e., particle filters and statistical linearization. Unfortunately, the naïve application of both particle filters and statistical linearization is bound to fail. In the remainder of this thesis, statistical linearization techniques are further pursued; It is shown that the sophisticated application of statistical linearization allows for deriving closed-form expressions for many relevant extended object tracking problems.

CHAPTER 3

Conic Fitting using Statistical Linearization

This chapter is devoted to a particular extended object tracking problem in which the measurements originate from a closed curve. In this case, the problem can be interpreted as a "dynamic" curve fitting problem for which a Bayesian state estimator is desired. We focus on conics such as circles and ellipses, which are highly relevant for practical applications. For example, when tracking people with a laser rangefinder, it is suitable to model the people as ellipses, where reflections are obtained from the ellipse boundary.

Conic fitting is typically formulated as an implicit problem, i.e., the conic is represented as an implicit equation and each measurement originates from an unknown measurement source that satisfies the implicit equation. In order to render the implicit problem explicit, existing Gaussian-based approaches perform a linearization around both the measurement and state. Thus, they suffer from a poor estimation quality in case the measurement noise is large in compared to the conic. As a consequence, these approaches are unsuitable for extended object tracking, where one often has to deal with rather large measurement noise.

In this chapter, we propose a new approach tailored to the requirements of extended object tracking. The basic idea is to approximate the original implicit problem with an explicit measurement equation that is corrupted with multiplicative noise.

Based on the explicit measurement equation, a Gaussian filter that performs a statistical linearization is employed for a recursive closed-form measurement update. Especially for large measurement noise, the novel Gaussian filter may significantly outperform state-of-the-art Gaussian filters.

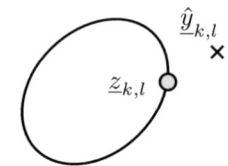

Remark 3.1. This chapter is based on [13,16], i.e., the circle fitting method has been published in [16] and the general method for conics has been published in [13]. This chapter

Figure 3.1: Illustration of the measurement source $\underline{z}_{k,l}$ and the measurement $\hat{\underline{y}}_{k,l}$ for an ellipse.

extends [13, 16] by an exhaustive analysis and comparison of different measurement equations and their corresponding likelihood functions.

3.1 Problem Formulation

In this chapter, we consider a special subclass of the general problem stated in Section 2.1 in which the measurement sources come from a conic such as an ellipse (without its interior).

In this case, the time-varying parameter vector of a conic specified by the state vector $\underline{x}_k = \left[\underline{m}_k^T, (\underline{x}_k^*)^T, \underline{p}_k^T\right]^T$ is to be estimated, where $\underline{m}_k$ is the center, $\underline{x}_k^*$ summarizes further kinematic parameters, and $\underline{p}_k$ are the conic parameters.

A conic is a two-dimensional quadric and a quadric is the solution of an implicit quadratic equation, i.e.,

$$\mathcal{M}(\underline{x}_k) := \{\underline{z} \in \mathbb{R}^N \mid g(\underline{x}_k, \underline{z}) = 0\} \ , \tag{3.1}$$

where $g(\underline{x}_k, \underline{z})$ is of the form

$$\underline{z}^T \mathbf{A}_k \underline{z} + \underline{b}_k^T \underline{z} + f_k \ , \tag{3.2}$$

with $N \times N$-dimensional symmetric matrix $\mathbf{A}_k$, N-dimensional vector $\underline{b}_k$, and constant f_k. The state vector $\underline{x}_k$ is a suitable parameterization of $\mathbf{A}_k$, $\underline{b}_k$, and f_k.

Example 3.1. A conic section can be parameterized by a six-dimensional vector $\underline{p}_k^{co} := [a_k, b_k, c_k, d_k, e_k, f_k]^T$, which means that $\mathbf{A}_k := \begin{bmatrix} a_k & c_k \\ c_k & b_k \end{bmatrix}$, and $\underline{b}_k := [d_k, e_k]^T$. Note that the center is implicitly encoded in the parameters $\underline{p}_k^{co}$.

Measurement Model   At each time step, a set of n_k measurements $\hat{\mathcal{Y}}_k = \{\hat{\underline{y}}_{k,l}\}_{l=1}^{n_k}$ is available. Each measurement $\hat{\underline{y}}_{k,l}$ is a noisy observation of an unknown measurement source $\underline{z}_{k,l}$ according to

$$\hat{\underline{y}}_{k,l} = \underline{z}_{k,l} + \underline{v}_{k,l} \ , \tag{3.3}$$

where $\underline{v}_{k,l}$ is zero-mean white Gaussian noise with known covariance matrix $\Sigma_{k,l}^v$ (see Fig. 3.1). The measurement source $\underline{z}_{k,l}$ lies on the conic, i.e.,

$$g(\underline{x}_k, \underline{z}_{k,l}) = 0 \ . \tag{3.4}$$

Hence, the *extent model* for conics just assumes that the measurement source is an element of the conic. No further information about the measurement source $\underline{z}_{k,l}$ on the conic section is available. In particular, no statistical information, e.g., a probability distribution for $\underline{z}_{k,l}$, is available. However, of course, specific inference algorithms may impose further assumptions on the measurement sources (e.g. a uniform distribution).

Remark 3.2. Equation (3.3) and (3.4) specify an *implicit* measurement equation, which can be interpreted as an implicit Errors-In-Variables (EIV) Model [Che10].

Dynamic Model   The temporal evolution of the state vector $\underline{x}_k$ may be given by a stochastic dynamic model as discussed previously in Section 2.1.3.

3.2   Related Work

Conic fitting, or more general curve fitting, is a traditional, well-studied problem. However, in contrast to extended object tracking, curve fitting is usually understood to be a static problem and the measurement noise is isotropic, where its statistics are unknown.

In this section, a brief overview of the most well-known conic fitting methods is given. Due to the tracking background, we mostly focus on recursive Bayesian state estimators as they allow to incorporate a probabilistic motion model and a sound treatment of stochastic measurement noise whose statistics may vary from measurement to measurement. In particular, the *Extended Kalman Filter (EKF)* approach [Por90, EAB92, Ros93, Zha97] to conic fitting is discussed in detail as it will be compared to the novel approach later in this chapter.

Deterministic Methods   Deterministic fitting methods do not make use of a statistical description of the problem. As these methods do not directly allow for incorporating probabilistic information about the measurement noise or the temporal evaluation of the object, they are in general unsuitable for extended object tracking.

A popular deterministic method for curve fitting is *orthogonal distance regression* [Che10, CL05] that aims at minimizing the orthogonal geometric distance from the measurements to the curve. As this is a nonlinear least-squares problem, no closed-form solutions can be derived in general. Even for simple basic curves, iterative optimization methods such as the Gauss-Newton method or the Levenberg-Marquardt algorithm [Che10] have to be used to find an approximate solution.

A computationally more attractive alternative to the geometric fit is the so-called *algebraic fit*, which minimizes a simpler distance measure. For instance, the *Delogne-Kåsa method* [Che10] is a widely-used algebraic fit that yields a linear least-squares problem. In general, algebraic fits are less accurate than geometric fits. However, geometric fits are computationally far more expensive and may stuck in a local minimum.

Statistical Methods   Essentially, there are two kinds of statistical formulations of the curve fitting problem [Che10, ASC09, Cha65, Kan96]. In the *functional model*, the measurement sources are assumed to be fixed but unobservable, it is only known that they lie on the curve. In this case, the measurement sources are treated as additional model parameters that are to be estimated. In the *structural model*, the measurement sources are independent realizations of a random variable with a probability distribution concentrated on the curve. Usually, the parameters of this distribution are estimated as well.

As pointed out in [Cha65], the *Maximum Likelihood Estimator (MLE)* based on the functional model coincides with the geometric fit for circles. However, the MLE based on the structural model in general differs from the geometric fit. The statistical models allow for a statistical error analysis of deterministic fitting methods, e.g., the bias can be determined (see [Che10, ASC09] for an analysis of circle fits).

Bayesian methods for curve fitting can be found for example in [Por90, EAB92, Ros93, NR97, WK01]. In case the measurement sources are not estimated explicitly, the basic approach is to assume the measurement sources to be drawn uniformly distributed from the curve [WK01]. As the likelihood cannot be computed analytically, elaborate approximation techniques such as particle filters are required for a recursive update.

There are some approaches that do not use an explicit likelihood for the measurement sources. Instead these approaches are based on a linearization of the implicit shape function (3.4) around both the measurement $\hat{\underline{y}}_{k,l}$ and the (current) estimate of the state $\underline{x}_k^*$ [Por90, EAB92, Ros93, Zha97], i.e.,

$$g(\underline{x}_k, \underline{z}_{k,l}) \approx g(\underline{x}_k^*, \hat{\underline{y}}_{k,l}) + \mathbf{H}_{k,l}^x \left(\underline{x}_k - \underline{x}_k^* \right) + \mathbf{H}_{k,l}^z \left(\underline{z}_{k,l} - \hat{\underline{y}}_{k,l} \right) \quad (3.5)$$

with

$$\mathbf{H}_{k,l}^x = \frac{\partial g}{\partial \underline{x}_k}(\underline{x}_k^*, \hat{\underline{y}}_{k,l}) \text{ , and}$$

$$\mathbf{H}_{k,l}^z = \frac{\partial g}{\partial \underline{z}_{k,l}}(\underline{x}_k^*, \hat{\underline{y}}_{k,l}) \text{ .}$$

Based on the linearized implicit equation, the following explicit linear measurement equation is obtained

$$\hat{\underline{y}}_{k,l}^* = \mathbf{H}_{k,l}^x \, \underline{x}_k + \underline{v}_{k,l}^* \quad (3.6)$$

with

$$\hat{\underline{y}}_{k,l}^* := -g(\underline{x}_k^*, \hat{\underline{y}}_{k,l}) + \mathbf{H}_{k,l}^x \, \underline{x}_k^* \text{ ,}$$

$$\underline{v}_{k,l}^* := \mathbf{H}_{k,l}^z \, \underline{v}_{k,l} \text{ .}$$

As the above system is linear, the standard Kalman filter [Sim06] can be used for the measurement update.

It is important to note that in contrast to standard the *Extended Kalman Filter (EKF)* for explicit equations, the linearization is also performed around the measurement. The downside of the linearization around the measurement is that poor estimation results may be obtained in case of high measurement noise (see also the evaluation in Section 3.4.4). As a consequence, this approach is rather unsuitable for extended object tracking problems, where one typically has to deal with large noise. Additionally, the *EKF* approaches to ellipse fitting [Por90, EAB92, Ros93] employ an ellipse parameterization that does not explicitly contains the center of the ellipse. Rewriting the linear motion models from Section 2.1.3 for this parameterization results in a nonlinear system equation.

3.3 General Approach

In the following, an explicit measurement equation based on the implicit problem (see Section 3.1) is derived for general conics. The two subsequent sections are devoted to two special conics, i.e., circles and ellipses.

The basic idea is to substitute both hand-sides of (3.3) in (3.4)

$$g(\underline{x}_k, \hat{\underline{y}}_{k,l}) = g(\underline{x}_k, \underline{z}_{k,l} + \underline{v}_{k,l}) , \qquad (3.7)$$

and to separate the measurement source $\underline{z}_{k,l}$ from the measurement noise $\underline{v}_{k,l}$ by performing the following algebraic reformulation

$$g(\underline{x}_k, \hat{\underline{y}}_{k,l}) = \underbrace{g(\underline{x}_k, \underline{z}_{k,l})}_{=0} + g^v(\underline{x}_k, \underline{v}_{k,l}, \underline{z}_{k,l}) , \qquad (3.8)$$

where $g^v(\underline{x}_k, \underline{v}_{k,l}, \underline{z}_{k,l}) = 2\underline{z}_{k,l}^T \mathbf{A}_k \underline{v}_{k,l} + \underline{v}_{k,l}^T \mathbf{A}_k \underline{v}_{k,l} + \underline{b}_k^T \underline{v}_{k,l}$.

Based on (3.8), we can construct the final measurement equation

$$0 = g(\underline{x}_k, \hat{\underline{y}}_{k,l}) - g^v(\underline{x}_k, \underline{v}_{k,l}, \underline{z}_{k,l}) \qquad (3.9)$$

$$=: h(\underline{x}_k, \underline{v}_{k,l}, \hat{\underline{y}}_{k,l}) , \qquad (3.10)$$

where $h(\underline{p}_k, \underline{v}_{k,l}, \hat{\underline{y}}_{k,l})$ is a measurement function that maps the random variables $\underline{x}_k$ and $\underline{v}_{k,l}$ to a pseudo-measurement with value 0.

The unknown measurement source $\underline{z}_{k,l}$ in (3.9) can be substituted with a proper point estimate based on the latest conic estimate. The introduced error through the point estimate is minor because $\mathrm{E}\{g^v(\underline{x}_k, \underline{v}_{k,l}, \underline{z}_{k,l})\}$ is independent of $\underline{z}_{k,l}$ due to the zero mean noise $\underline{v}_{k,l}$.

Measurement equation (3.9) is a quadratic equation corrupted with Gaussian distributed *multiplicative* noise. We aim at a Gaussian filter via statistical linearization of the measurement equation (3.9). Hence, the posterior probability density for the state $\underline{x}_k$ having incorporated the first $l-1$ measurements from time step k is given by $f_{l-1}(\underline{x}_k) = \mathcal{N}(\underline{x}_k - \underline{\mu}^x_{k,l-1}, \Sigma^x_{k,l-1})$. Statistical linearization according to the procedure described in Section 2.4 leads to the updated density $f_l(\underline{x}_k) = \mathcal{N}(\underline{x}_k - \underline{\mu}^x_{k,l}, \Sigma^x_{k,l})$ with the measurement $\hat{\underline{y}}_{k,l}$

$$
\begin{aligned}
\underline{\mu}^x_{k,l} &= \underline{\mu}^x_{k,l-1} + \Sigma^{xh}_{k,l} \left(\Sigma^{hh}_{k,l}\right)^{-1} \left(0 - \mu^h_{k,l}\right) \ , \\
\Sigma^x_{k,l} &= \Sigma^x_{k,l-1} - \Sigma^{xh}_{k,l} \left(\Sigma^{hh}_{k,l}\right)^{-1} \Sigma^{hx}_{k,l} \ ,
\end{aligned}
$$

where $\mu^h_{k,l}$ is the predicted pseudo-measurement, $\Sigma^{xh}_{k,l}$ is the covariance between the pseudo-measurement and state, and $\Sigma^{hh}_{k,l}$ is the variance of the predicted pseudo-measurement. The required moments can be calculated analytically according to Section 2.4.1.

Discussion The above described approach can be interpreted as a statistical linearization around the *measurement source* $\underline{z}_{k,l}$ and the state $\underline{x}_k$ with regards to the algebraic properties of the shape function (see Section 2.4.1). As the measurement source $\underline{z}_{k,l}$ is unknown, an explicit linearization would be rather unsuitable, however, the statistical linearization can be performed without (or with little) knowledge about the measurement sources. Please note that if the function $g(\underline{x}_k, \underline{z}_{k,l})$ would be linear in both $\underline{x}_k$ and $\underline{z}_{k,l}$, measurement equation (3.7) and the explicit linearization (3.6) coincide. The proposed Gaussian filter minimizes the (statistically linearized) *algebraic distance* between the measurements and the conic, where the statistics of the noise is incorporated.

3.4 Special Case: Circle Fitting

In this section, we apply the general approach described in the previous section to circles. The problem of fitting a circle to noisy data points occurs in many application areas such as computer vision [FB08], physics [Kar91], and medicine [SSB+07]. An example in the context of extended object tracking is the tracking of circle-shaped landmarks [NVMBS08, ZXA06] in mobile robotics.

A circle can be specified by its center and radius, i.e., the parameter vector for a circle $\boldsymbol{p}_k^{\text{ci}} = \boldsymbol{r}_k$ only consists of radius $\boldsymbol{r}_k$. The implicit shape function (see (3.4)) for a circle is of the form

$$g^{\text{ci}}(\underline{\boldsymbol{x}}_k^{\text{ci}}, \underline{\boldsymbol{z}}_{k,l} - \underline{\boldsymbol{m}}_k) = ||\underline{\boldsymbol{z}}_{k,l} - \underline{\boldsymbol{m}}_k||^2 - \boldsymbol{r}_k^2 \qquad (3.11)$$

with $\underline{\boldsymbol{x}}_k^{\text{ci}} = \left[\underline{\boldsymbol{m}}_k^T, (\underline{\boldsymbol{x}}_k^*)^T, \boldsymbol{p}_k^{\text{ci}}\right]^T$.

In the following, the version of (3.9) for circles is compared with two further reasonable measurement equations in order to obtain further insights, e.g., the corresponding likelihood functions are computed. All in all, we investigate the following three measurement equations.

- *Polar Equation* (Section 3.4.1)
  When the measurement sources are assumed to be uniformly distributed on the circle, an obvious explicit measurement equation can be obtained when rewriting the problem to polar coordinates. However, it turns out that this equation is unsuitable for statistical linearization because it is unclear how to compute the required moments analytically and the Gaussian approximation is poor. Hence, the naïve statistical linearization for circle fitting is not promising.

- *Explicit Measurement Equation 1* (Section 3.4.2)
  The measurement equation according to the basic approach described Section 3.3, i.e., (3.9) for circles. In case of isotropic noise, this equation is equivalent to the polar equation (3.12), i.e., the corresponding likelihoods are equal.

- *Explicit Measurement Equation 2* (Section 3.4.3)
  An alternative measurement equation that is equivalent to the polar equation, which however, requires the statistical linearization around the measurement.

In fact, it emerges that the *explicit measurement equation 1* according to Section 3.3 is most suitable for the considered application as it outperforms all other equations in case of high measurement noise. At this point, it is essential to note that even when two measurement equations specify the same likelihood function, statistical linearization may cause that the resulting estimation quality significantly differs.

3.4.1 Polar Equation

Equation (3.11) can be written in (explicit) polar form

$$\underline{\hat{y}}_{k,l} = \underbrace{r_k \cdot \underline{e}(\phi_{k,l}) + \underline{m}_k}_{\underline{z}_{k,l}} + \underline{v}_{k,l} \ , \tag{3.12}$$

$$=: h_1^{\mathrm{ci}}(\underline{x}_k^{\mathrm{ci}}, \underline{v}_{k,l}, \phi_{k,l}) \ , \tag{3.13}$$

where $\phi_{k,l} \in [0, 2\pi]$ is an *unknown angle* and $\underline{e}(\phi_{k,l}) := [\cos(\phi_{k,l}), \sin(\phi_{k,l})]^T$ is a unit vector in direction $\phi_{k,l}$. When the measurement source is uniformly distribution on the circle [NR97], the angle $\phi_{k,l}$ is uniformly distributed on the interval $[0, 2\pi]$. In this case, the likelihood function specified by (3.13) becomes

$$
\begin{aligned}
f_u^L(\underline{x}_k^{\mathrm{ci}}) &= \int f(\underline{\hat{y}}_{k,l} \,|\, \underline{z}_{k,l}) \cdot p(\underline{z}_{k,l} \,|\, \underline{x}_k^{\mathrm{ci}}) \, \mathrm{d}\underline{z}_{k,l} \\
&= \frac{1}{2\pi r_k} \int_{\mathbf{K}(\underline{x}_k^{\mathrm{ci}})} p(\underline{\hat{y}}_{k,l} \,|\, \underline{z}_{k,l}) \, \mathrm{d}\underline{z}_{k,l} \\
&= \frac{1}{2\pi} \int f(\underline{\hat{y}}_{k,l} \,|\, \phi_{k,l}, \underline{x}_k^{\mathrm{ci}}) \, \mathrm{d}\phi_{k,l} \ ,
\end{aligned}
\tag{3.14}
$$

where $\mathbf{K}(\underline{x}_k^{\mathrm{ci}}) = \mathbf{K}(\underline{m}_k, r_k) := \{\underline{z} \mid \underline{z} \in \mathbb{R}^2 \text{ and } \|\underline{z} - \underline{m}_k\|_2 \le r_k\}$ is the circle with center $\underline{m}_k$ and radius r_k.

In general, this likelihood function cannot be solved analytically. However, for isotropic Gaussian measurement noise, it is possible to derive explicit expressions in terms of the non-central χ^2-distribution.

Definition 3.1 (Non-central χ^2-distribution). The non-central χ^2-distribution with κ degrees of freedom and non-centrality parameter λ is defined as

$$f_{\chi^2}^{\kappa, \lambda}(x) = \frac{1}{2} e^{-(x+\lambda)/2} \left(\frac{x}{\lambda}\right)^{\kappa/4 - 1/2} I_{\kappa/2 - 1}\left(\sqrt{\lambda x}\right) \ ,$$

where $I_\nu(z)$ denotes the modified *Bessel function* of first kind

$$I_\nu(y) = (y/2)^\nu \sum_{j=0}^{\infty} \frac{(y^2/4)^j}{j!\Gamma(\nu+j+1)} \quad .$$

Theorem 3.1. *For isotropic Gaussian noise $\underline{v}_{k,l}$ with covariance matrix $\Sigma_{k,l}^v = \mathrm{diag}(\sigma^2,\sigma^2)$, the likelihood function $f_u^L(\underline{x}_k^{ci})$ in (3.14) is given by*

$$f_u^L(\underline{x}_k^{ci}) \sim f_{\chi^2}^{\kappa,\lambda}\left(\frac{1}{\sigma^2}r_k^2\right) \quad ,$$

where $f_{\chi^2}^{\kappa,\lambda}(\cdot)$ denotes the density of the non-central χ^2-distribution with $\kappa = 2$ degrees of freedom and non-centrality parameter $\lambda = \frac{1}{\sigma^2} \cdot ||\underline{\hat{y}}_{k,l} - \underline{m}_k||^2$.

Proof. According to [NR97], the integral in (3.14) can be evaluated analytically as

$$f_u^L(\underline{x}_k^{ci}) = \frac{1}{2\pi\sigma^2} \frac{r_k}{|\mathbf{K}(\underline{m}_k,r_k)|} \cdot \exp\left\{ -\frac{(r_k - ||\underline{\hat{y}}_{k,l} - \underline{m}_k||)^2}{2\sigma^2} \right\}$$

$$\cdot \exp\left\{ -\frac{r_k ||\underline{\hat{y}}_{k,l} - \underline{m}_k||}{\sigma^2} \right\} \cdot I_0\left(\frac{r_k ||\underline{\hat{y}}_{k,l} - \underline{m}_k||}{\sigma^2} \right) \quad , \quad (3.15)$$

where $|\mathbf{K}(\underline{m}_k, r_k)| = 2\pi r_k$ is the circumference of a circle with radius r_k. With the definition of a non-central χ^2-distribution, we obtain

$$f_u^L(\underline{x}_k^{ci}) \sim \frac{1}{2} \exp\left\{ -\frac{r_k^2 + ||\underline{\hat{y}}_{k,l} - \underline{m}_k||^2}{2\sigma^2} \right\} \cdot I_0\left(\frac{r_k ||\underline{\hat{y}}_{k,l} - \underline{m}_k||^2}{\sigma^2} \right) \quad (3.16)$$

$$= f_{\chi^2}^{\kappa,\lambda}\left(\frac{1}{\sigma^2} \cdot r_k^2 \right) \quad (3.17)$$

with $\lambda = \frac{1}{\sigma^2} \cdot ||\underline{\hat{y}}_{k,l} - \underline{m}_k||^2$. $\qquad\square$

As already mentioned, the polar equation as given by (3.13) is rather unsuitable for a Gaussian filter based on statistical linearization (see Section 2.4). The reason is that a Gaussian filter approximates the measurement $h_1^{ci}(\underline{x}_k, \underline{v}_{k,l}, \phi_{k,l})$ with a Gaussian distribution whose mean coincides with the center of the circle, which is a poor approximation of the predicted measurement. Second, it remains open how to calculate the moments required for a Gaussian state estimator in closed form.

3.4.2 Explicit Measurement Equation 1

In the following, we investigate the measurement equation for circle fitting following the novel approach presented in Section 3.3. Substituting the measurement $\underline{\hat{y}}_{k,l}$ (3.3) into the implicit circle equation (3.11) results in

$$g^{\text{ci}}(\underline{x}_k^{\text{ci}}, \underline{\hat{y}}_{k,l}) \;=\; g^{\text{ci}}(\underline{x}_k^{\text{ci}}, \underline{z}_{k,l} + \underline{v}_{k,l}) \;, \tag{3.18}$$

which can be rewritten as

$$||\underline{\hat{y}}_{k,l} - \underline{m}_k||^2 - r_k \;=\; ||\underline{z}_{k,l} + \underline{v}_{k,l} - \underline{m}_k||^2 - r_k \tag{3.19}$$

$$\;=\; 2\langle \underline{z}_{k,l}, \underline{v}_{k,l} \rangle + ||\underline{v}_{k,l}||^2 \;. \tag{3.20}$$

With $\underline{z}_{k,l} = r_k \cdot \underline{e}(\phi_{k,i})$, we obtain the explicit measurement equation

$$0 \;-\; ||\underline{\hat{y}}_{k,l} \quad \underline{m}_k||^2 - r_k^2 + 2 \cdot r_k \cdot \langle \underline{e}(\phi_{k,l}), \underline{v}_{k,l} \rangle + ||\underline{v}_{k,l}||^2 \tag{3.21}$$

$$=: \; h_1^{\text{ci}}(\underline{x}_k, \underline{v}_{k,l}, \underline{\hat{y}}_{k,l}) \;, \tag{3.22}$$

where $h_1^{\text{ci}}(\cdot)$ maps the state vector $\underline{x}_k$, the measurement noise $\underline{v}_{k,l}$, and the measurement $\underline{\hat{y}}_{k,l}$ to a pseudo-measurement 0. Note that this measurement equation still depends on the unknown angle $\phi_{k,l}$. However, it turns out that for isotropic $\underline{v}_{k,l}$, the resulting likelihood does not depend on $\phi_{k,l}$ and the likelihood is the same as assuming a uniform distribution for the measurement source.

Theorem 3.2. *For isotropic Gaussian noise $\underline{v}_{k,l}$ with covariance matrix $\Sigma_{k,l}^v = \text{diag}(\sigma^2, \sigma^2)$, the likelihood function $f_1^L(\underline{x}_k^{ci})$ specified by (3.21) is given by*

$$f_1^L(\underline{x}_k^{ci}) \sim f_{\chi^2}^{\kappa, \lambda}\left(\frac{1}{\sigma^2} ||\underline{\hat{y}}_{k,l} - \underline{m}_k||^2 \right) \;, \tag{3.23}$$

where $f_{\chi^2}^{\kappa, \lambda}(\cdot)$ denotes the density of the non-central χ^2-distribution with $\kappa = 2$ degrees of freedom and non-centrality parameter $\lambda = \frac{1}{\sigma^2} \cdot r_k^2$.

Proof. For given $\underline{m}_k$, the term $\frac{1}{\sigma^2}||\underline{z}_{k,l} + \underline{v}_{k,l} - \underline{m}_k||^2$ in (3.19) is distributed according to the non-central χ^2-distribution with $\kappa = 2$ and $\lambda = \frac{1}{\sigma^2} \cdot r_k^2$. $\square$

Remark 3.3. For non-isotropic noise, we obtain a generalized χ^2-distribution that depends on $\phi_{k,l}$.

Theorem 3.3. *In two-dimensional space, the likelihood* (3.23) *is the same as* (3.14) *for isotropic Gaussian noise* $\underline{v}_{k,l}$, *i.e.,* $f_1^L(\underline{x}_k^{ci}) = f_u^L(\underline{x}_k^{ci})$.

Proof. Follows from the definition of the non-central χ^2-distribution with $\kappa = 2$ degrees of freedom:

$$f_{\chi^2}^{\kappa,\lambda}(x) = \frac{1}{2} \exp\left\{-(x+\lambda)/2\right\} \cdot I_0\left(\sqrt{\lambda x}\right) \quad.$$

$\square$

Measurement equation (3.21) has an intuitive interpretation. It considers the squared distance from the measurement to the center $||\underline{\hat{y}}_{k,l} - \underline{m}_k||^2$ and determines how likely it is to obtain a measurement with this distance to the circle. With this interpretation in mind, it becomes clear that the likelihood is independent of $\phi_{k,l}$ in case of isotropic noise as the angle does not influence the probability distribution of the distance.

For given center and radius, the predicted pseudo-measurement is a non-central χ^2-square distribution, which is uni-modal and can be approximated well with a Gaussian distribution.

A further mentionable insight is that for given center $\underline{\hat{m}}_k$ the radius can be estimated *independently* of the true measurement sources due to

$$\underbrace{||\underline{\hat{y}}_{k,l} - \underline{\hat{m}}_k||_k^2}_{:=\underline{\hat{y}}_{k,l}^*} \;=\; \underbrace{r_k^2 + 2 \cdot r_k \cdot \langle \underline{e}_k(\phi_{k,l}), \underline{v}_{k,l}\rangle + ||\underline{v}_{k,l}||^2}_{h_1^{ci,m}(r_k, \underline{v}_{k,l})} \quad. \quad (3.24)$$

is a usual explicit measurement equation, which maps the state r_k and the noise $\underline{v}_{k,l}$ to a pseudo-measurement $\underline{\hat{y}}_{k,l}^*$.

Remark 3.4. In case of non-isotropic noise, we suggest to substitute the unknown angle $\phi_{k,l}$ with a point estimate, which then serves as the linearization point. Then, the likelihood specified by (3.21) can also be interpreted as an approximation of (3.14). Later in Section 3.4.4, we demonstrate with simulations that the resulting approximation error is rather negligible.

3.4.3　Explicit Measurement Equation 2

Instead of directly inserting the measurement into (3.11), it is also reasonable to insert the expression $\underline{\hat{y}}_{k,l} - \underline{v}_{k,l}$ for the measurement source into

(3.11), i.e.,

$$g^{\text{ci}}(\underline{x}_k^{\text{ci}}, \hat{\underline{y}}_{k,l} - \underline{v}_{k,l}) = 0 , \tag{3.25}$$

which results in the measurement equation

$$0 = ||\hat{\underline{y}}_{k,l} - \underline{v}_{k,l} - \underline{m}_k||^2 - r_k^2 \tag{3.26}$$

$$=: h_2^{\text{ci}}(\underline{x}_k^{\text{ci}}, \underline{v}_{k,l}, \hat{\underline{y}}_{k,l}) . \tag{3.27}$$

Actually, the likelihood specified by this measurement equation also coincides with the likelihood that results from uniformly distributed measurement sources.

Theorem 3.4. *For isotropic Gaussian noise $\underline{v}_{k,l}$ with covariance matrix $\Sigma_{k,l}^v = \text{diag}(\sigma^2, \sigma^2)$, the likelihood function $f_2^L(\underline{x}_k^{ci})$ specified by (3.26) is given by*

$$f_2^L(\underline{x}_k^{ci}) \sim f_{\chi^2}^{\kappa,\lambda}\left(\frac{1}{\sigma^2} r_k^2\right) , \tag{3.28}$$

where $f_{\chi^2}^{\kappa,\lambda}(\cdot)$ denotes the density of the non-central χ^2-distribution with $\kappa = 2$ degrees of freedom and non-centrality parameter $\lambda = \frac{1}{\sigma^2} \cdot ||\hat{\underline{y}}_{k,l} - \underline{m}_k||^2$.

Proof. For given $\underline{m}_k$, the term $\frac{1}{\sigma^2} \cdot ||\hat{\underline{y}}_{k,l} - \underline{v}_{k,l} - \underline{m}_k||^2$ in (3.19) is distributed according to the non-central χ^2-distribution with $\kappa = 2$ and $\lambda = \frac{1}{\sigma^2} \cdot ||\hat{\underline{y}}_{k,l} - \underline{m}_k||^2$. □

Remark 3.5. The likelihood function $f_2^L(\underline{x}_k^{ci})$ specified by (3.26) in general coincides with the likelihood (3.14), i.e., $f_1^L(\underline{x}_k^{ci}) \sim f_u^L(\underline{x}_k^{ci})$.

Like (3.21), (3.26) is a polynomial equation for which the uniformly distributed noise term has been removed. The measurement equation (3.26) can be directly statistically linearized as described in Section 2.4. However, as the statistical linearization is performed around the measurement, a large linearization error is expected to occur for high measurement noise.

3.4.4 Evaluation

In this section, we provide a detailed evaluation of the derived measurement equations when employing statistical linearization. The measurements for performing the circle fits arise from a *stationary* circle with center $[5,5]^T$ and radius 3, i.e., the circle neither moves nor changes its radius.

Remark 3.6. The evaluation is restricted to a stationary circle as all methods are equal for the time update step. They all employ the same parameterization of a circle so that the same motion models can be used.

Two scenarios are considered, which differ in the set of possible measurement sources as depicted in Fig. 3.2. In the first scenario, the measurement sources are equidistantly distributed on the circle. The second scenario covers a situation in which the circle is observed from a specific point of view, i.e., measurement sources are concentrated on the upper right segment of the circle. Measurements are received sequentially from the circle, i.e., at each time step k exactly one measurement is obtained ($n_k = 1$). The measurement source for a measurement is drawn uniformly from the set of possible measurement sources.

For each scenario, simulations are performed for different measurement noise levels, i.e.,

a) low isotropic noise $\Sigma_{k,1}^v = \mathrm{diag}(0.3, 0.3)$,

b) high isotropic noise $\Sigma_{k,1}^v = \mathrm{diag}(2, 2)$, and

c) non-isotropic noise $\Sigma_{k,1}^v = \mathrm{diag}(1, 0.4)$.

Measurement Equation 1 and 2 vs. EKF   First, we compare the Gaussian filters based on *statistical linearization* of measurement equation 1 and 2 (using analytic moment calculation as described in Section 2.4.1) with the *Extended Kalman Filter* approach as described in (3.5) [Zha97,Por90]. The prior for the circle parameters is Gaussian with mean $[4, 4, 2.5]^T$ and covariance matrix $\mathrm{diag}(1.5, 1.5, 0.7)$. Figures 3.3, 3.4, and 3.5 depict the estimation results for the first scenario and Figures 3.7, 3.8, and 3.9 for the second scenario.

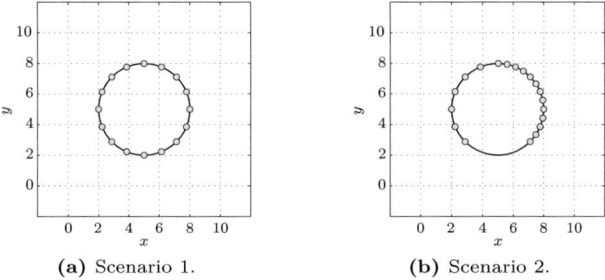

(a) Scenario 1.   (b) Scenario 2.

Figure 3.2: Considered scenarios with measurement sources.

The simulation results show that both measurement equation 1 and 2 significantly outperform the *EKF* solution (3.5). In particular, for large measurement noise the *EKF* solution yields poor estimation results. This result can be explained by the explicit linearization around the measurement and the current estimate, which suffers from large approximation errors when the noise is high. Even measurement equation 2 results in a small bias for the radius in case the measurement noise is large due to the statistical linearization around the measurement. All told, the simulations show that measurement equation 1 is most suitable for extended object tracking problems.

Measurement Equation 1 and 2 vs. Polar Equation   Second, we compare the Gaussian filters using statistical linearization of measurement equation 1 and 2 with the polar equation (3.12) (where a uniform assumption on the measurement sources is assumed). As there are no closed-form expression for the required moments available when using the polar equation, we employed Monte Carlo sampling in order to determine the moments approximately. The estimation results are depicted in Fig. 3.11. It can be seen that the Gaussian filter based on the polar equation gives significantly worse estimation results than the measurement equations 1 and 2, which can be explained by the poor Gaussian approximation when using the polar equation.

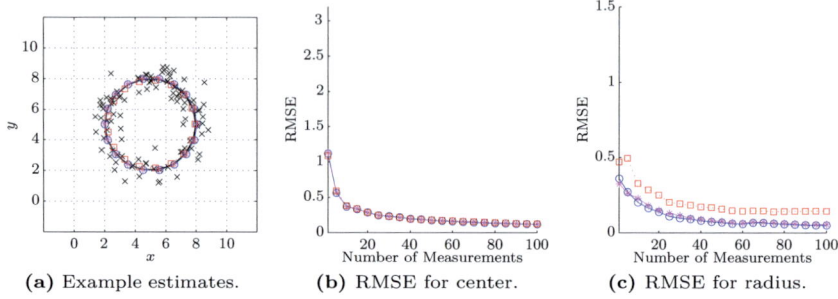

(a) Example estimates.   **(b)** RMSE for center.   **(c)** RMSE for radius.

Figure 3.3: Estimation results for Scenario 1a.

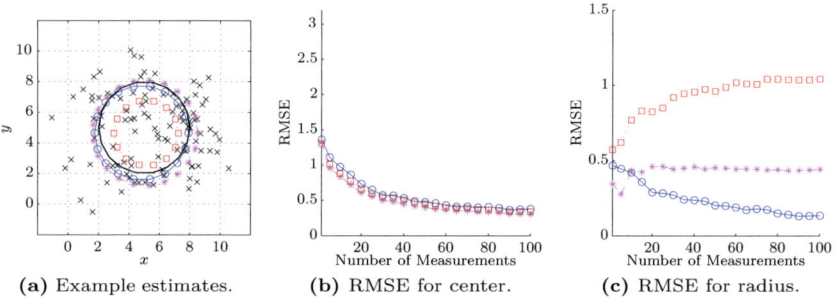

(a) Example estimates.   **(b)** RMSE for center.   **(c)** RMSE for radius.

Figure 3.4: Estimation results for Scenario 1b.

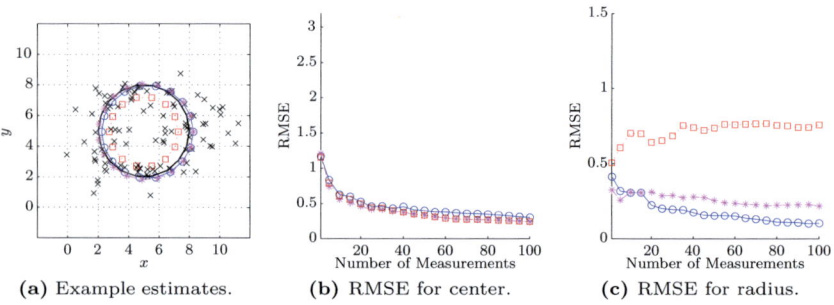

(a) Example estimates.   **(b)** RMSE for center.   **(c)** RMSE for radius.

Figure 3.5: Estimation results for Scenario 1c.

───◯───	Meas. Eqn. 1
─ ✳ ─	Meas. Eqn. 2
···□···	EKF

Figure 3.6: Legend for Scenario 1.

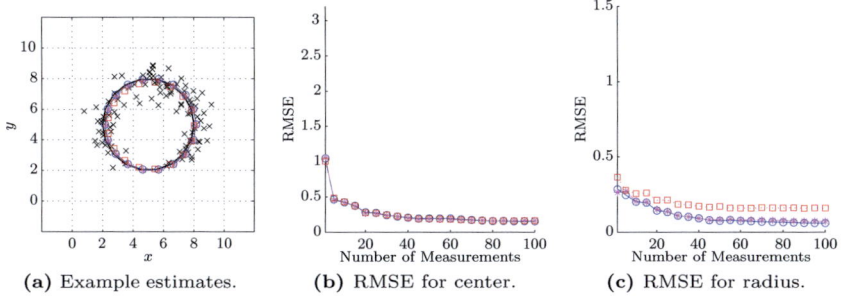

(a) Example estimates.   **(b)** RMSE for center.   **(c)** RMSE for radius.

Figure 3.7: Estimation results for Scenario 2a.

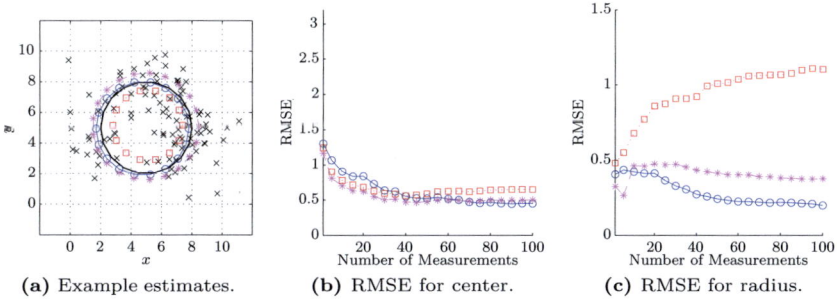

(a) Example estimates.   **(b)** RMSE for center.   **(c)** RMSE for radius.

Figure 3.8: Estimation results for Scenario 2b.

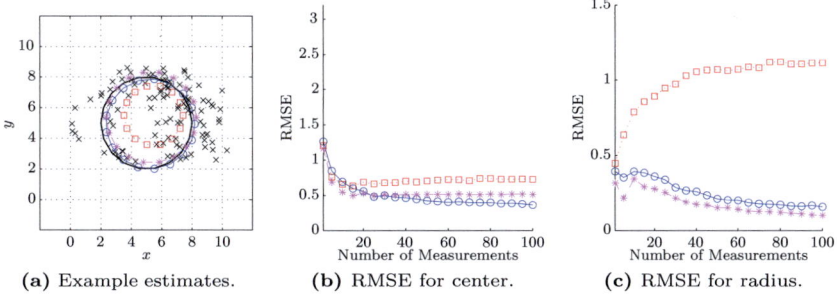

(a) Example estimates.   **(b)** RMSE for center.   **(c)** RMSE for radius.

Figure 3.9: Estimation results for Scenario 2c.

	Meas. Eqn. 1
	Meas. Eqn. 2
	EKF

Figure 3.10: Legend for Scenario 2.

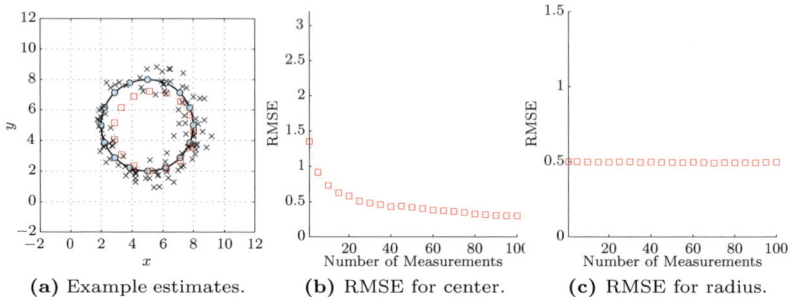

(a) Example estimates.   (b) RMSE for center.   (c) RMSE for radius.

Figure 3.11: Scenario 2b: Polar equation.

Discussion   The evaluation results show that both Gaussian filters based on measurement equation 1 and 2 significantly outperform the state-of-the-art Gaussian filters for circle fitting. As measurement equation 1 outperforms measurement equation 2 in case of high measurement noise, we prefer measurement equation 1 in the context of extended object tracking.

3.5   Special Case: Ellipse Fitting

Besides circles, the most relevant conics in the context of extended object tracking are ellipses. While the parameterization of a circle is rather obvious, the situation is different for an ellipse because the conic equation (3.2) only describes a valid ellipse if the matrix $\mathbf{A}_k$ is positive definite. Neglecting this constraint in an ellipse fitting procedure may decrease the fitting accuracy.

In the following, we present two ellipse parameterizations that employ the minimum number of five parameters in two-dimensional space. The first parameterization is well-known in computer vision, however, the second has not yet been used for ellipse fitting. For both parameterizations, we derive an explicit measurement equation as described in Section 3.3.

3.5.1 Explicit Measurement Equation: Coefficient Representation

According to [Por90, EAB92, Zha97], (3.2) can be expanded to

$$a_k x^2 + 2b_k xy + c_k y^2 + 2d_k x + 2e_k y + f_k = 0 \ ,$$

where $\underline{z} = [x, y]^T$. As the constraint $a_k + c_k \neq 0$ holds for an ellipse, the parameter c_k can be removed by assuming $a_k + c_k = 1$, i.e.,

$$g^{\text{coeff}}(\underline{x}_k^{\text{coeff}}, \underline{z}) := a_k(x^2 - y^2) + 2b_k xy + y^2 + 2d_k x + 2e_k y + f_k = 0 \quad (3.29)$$

with $\underline{x}_k^{\text{coeff}} = \left[(\underline{x}_k^*)^T, (\underline{p}_k^{\text{coeff}})^T \right]^T$ and parameter vector

$$\underline{p}_k^{\text{coeff}} = [a_k, b_k, d_k, e_k, f_k]^T \ .$$

Note that the center of the ellipse is implicitly encoded in $\underline{p}_k^{\text{coeff}}$. In the following, the above parameterization $\underline{p}_k^{\text{coeff}}$ is called coefficient representation of an ellipse.

Remark 3.7. An alternative normalization is to obey $f_k = 1$, which, however, excludes ellipses through the origin.

The coefficient representation comes with two disadvantages in the context of extended object tracking. First, not all parameters vectors $\underline{p}_k^{\text{coeff}}$ describe a valid ellipse. In computer vision, this is not always a serious problem as enough data is available to dismiss infeasible parameters. However, in extended object tracking, infeasible parameter estimates may frequently be obtained because only a few measurements per time instant are available and the ellipse parameters evolve over time. A second drawback of the coefficient representation is that the system model is usually formulated with regards of the ellipse center. However, the center is not directly contained in $\underline{p}_k^{\text{coeff}}$ and rewriting a system model in terms of the coefficient representation results in a nonlinear and unintuitive system equation.

The measurement equation (3.9) for the coefficient representation can be written as

$$
\begin{aligned}
0 &= g^{\text{coeff}}(\underline{x}_k^{\text{coeff}}, \hat{y}_k) - a_k(z_{k,l}^x v_{k,l}^x + (v_{k,l}^x)^2 - z_{k,l}^y v_{k,l}^y - (v_{k,l}^y)^2) \\
& \quad 2b_k(z_{k,l}^x v_{k,l}^y + z_{k,l}^y v_{k,l}^x + v_{k,l}^y v_{k,l}^x) \\
& \quad + z_{k,l}^y v_{k,l}^y + (v_{k,l}^y)^2 + 2d_k v_{k,l}^y + 2e_k v_{k,l}^x \\
&=: h^{\text{coeff}}(\underline{x}_k^{\text{coeff}}, \underline{v}_{k,l}) \; ,
\end{aligned}
\tag{3.30}
$$

where $\underline{z}_{k,l} = \left[z_{k,l}^x, z_{k,l}^y \right]^T$, $\underline{v}_{k,l} = \left[v_{k,l}^x, v_{k,l}^y \right]^T$, and $h^{\text{coeff}}(\underline{x}_k^{\text{coeff}}, \underline{v}_{k,l})$ is the measurement equation.

3.5.2   Explicit Measurement Equation: Center/Shape Representation

An N-dimensional ellipsoid with center $\underline{m}_k$ and positive definite shape matrix $\mathbf{B}_k$ is given by the solution of the implicit equation

$$
(\underline{z} - \underline{m}_k)^T \cdot \mathbf{B}_k^{-1} \cdot (\underline{z} - \underline{m}_k) - 1 = 0 \; ,
$$

where $\underline{z} \in \mathbb{R}^N$. In two-dimensional space, i.e., $N = 2$, the positive semi-definite matrix $\mathbf{B}_k^{-1}$ can be decomposed as $\mathbf{B}_k^{-1} = \mathbf{L}_k \mathbf{L}_k^T$, where

$$
\mathbf{L}_k := \begin{bmatrix} l_k^{(1)} & 0 \\ l_k^{(3)} & l_k^{(2)} \end{bmatrix}
\tag{3.31}
$$

is a lower triangular matrix (with positive diagonal entries). Hence, the parameter vector $\underline{p}_k^{\text{cs}}$ using the center/shape representation can be defined as $\underline{p}_k^{\text{cs}} := \left[l_k^{(1)}, l_k^{(2)}, l_k^{(3)} \right]^T$, which contains the center and the entries of the Cholesky decomposition. By this means, the quadratic function (3.1) can be written as

$$
g^{\text{cs}}(\underline{x}_k^{\text{cs}}, \underline{z}) := (\underline{z} - \underline{m}_k)^T \cdot (\mathbf{L}_k \mathbf{L}_k^T) \cdot (\underline{z} - \underline{m}_k) - 1
$$

with $\underline{x}_k^{\text{cs}} = \left[\underline{m}_k^T, (\underline{x}_k^*)^T, (\underline{p}_k^{\text{cs}})^T \right]^T$.

The state vector $\underline{x}_k^{\text{cs}}$ explicitly contains the ellipse center and each $\underline{p}_k^{\text{cs}}$ describes a valid ellipse as $\mathbf{L}_k \mathbf{L}_k^T$ is always positive definite. A further benefit is that this parameterization can directly be used for arbitrary dimensional ellipsoids.

For the center/shape parameterization, the measurement equation (3.9) becomes

$$
\begin{aligned}
0 &= g(\underline{x}_k^{\mathrm{cs}}, \hat{\underline{y}}_k) - 2(\underline{z}_{k,l} - \underline{m}_k)^T \mathbf{B}_k^{-1} \underline{v}_{k,l} + \underline{v}_k^T \mathbf{B}_k^{-1} \underline{v}_{k,l} \\
&=: h^{\mathrm{cs}}(\underline{x}_k^{\mathrm{cs}}, \underline{v}_{k,l}) \ ,
\end{aligned}
$$

where $h^{\mathrm{cs}}(\underline{x}_k^{\mathrm{cs}}, \underline{v}_{k,l})$ is the new measurement function. Note that we observed that the performance of a Gaussian estimator can be significantly improved, i.e., the linearization error becomes smaller, if the measurement equation is multiplied with the factor $\frac{1}{\mathrm{trace}\{\mathbf{L}_k \mathbf{L}_k^T\}}$.

3.5.3 Evaluation

The evaluation of the ellipse fitting methods is carried out by comparing

- a Gaussian filter based on statistically linearizing measurement equation (3.30) with

- the EKF approach [Por90, EAB92, Ros93] described in Section 3.2.

Note that both methods employ the coefficient representation of an ellipse discussed in Section 3.5 (the center/shape representation will be used in the next section).

We elaborate two scenarios with a stationary ellipse and different sets of possible measurement sources as shown in Fig. 3.12. Measurements are received sequentially from the ellipse and the measurement source for a measurement is drawn uniformly from the set of possible measurement sources. A priori of the ellipse parameters $\underline{x}_k^{\mathrm{coeff}}$ are initialized with a Gaussian with mean $[0.5, 0.5, 0, 0, -4.5]^T$ and covariance matrix $\mathrm{diag}(10, 10, 10, 10, 10)$, which corresponds to an uncertain circle with radius 1.5 located at the $[0.5, 0.5]^T$.

For each scenario, simulations are performed for two different measurement noise levels, i.e.,

  a) low isotropic noise $\Sigma_{k,l}^v = \mathrm{diag}(0.2, 0.2)$, and

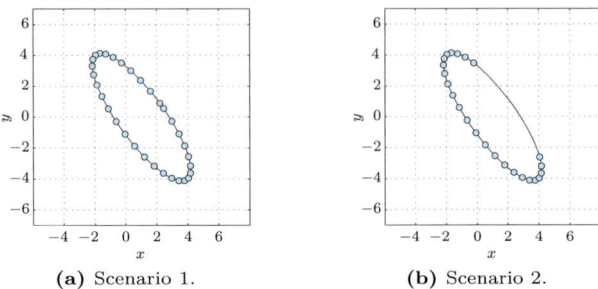

(a) Scenario 1.   (b) Scenario 2.

Figure 3.12: Considered scenarios: Ellipses with measurement sources.

b) high isotropic noise $\Sigma_{k,l}^{v} = \mathrm{diag}(0.5, 0.5)$.

Figures 3.13, and 3.14 depict the estimation results for Scenario 1 and the results for Scenario 2 are shown in Figures 3.16, and 3.17. The results demonstrate that the EKF solution is significantly outperformed by the new approach. In particular for large noise, the EKF suffers even more from the linearization error than in the case of circles.

3.6 Conclusions

This chapter was about the tracking of an extended object that is modeled as a conic such as a circle or an ellipse (without interior). As we assume that noisy point measurements from the conic are available, the estimation of the conic parameters can also be interpreted as a (statistical) curve fitting problem. Conic fitting is a traditional problem that has been extensively investigated in different areas such as computer vision or physics. However, in the context of extended object tracking, new requirements are imposed on the fitting algorithm. First, the estimator, i.e., the fitting algorithm, must be able to incorporate the conic's motion with the help of a probabilistic motion model. Second, the estimator must deal with few measurements whose noise is typically rather large compared to the size of the conic. As existing conic fitting methods are not tailored to these requirements, they are rather unsuitable for extended object tracking scenarios.

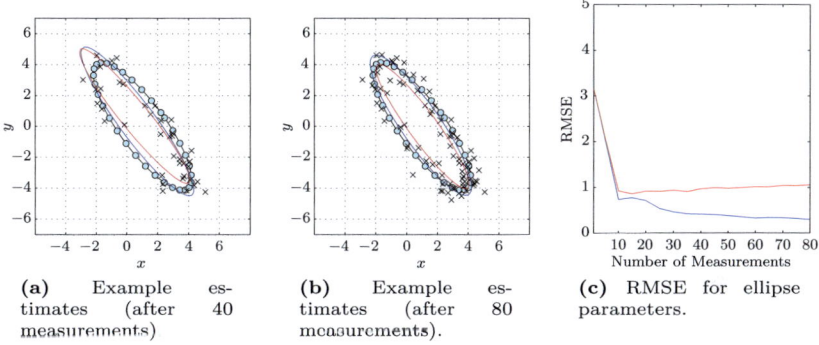

(a) Example es-
timates (after 40
measurements)

(b) Example es-
timates (after 80
measurements).

(c) RMSE for ellipse
parameters.

Figure 3.13: Estimation results for Scenario 1a.

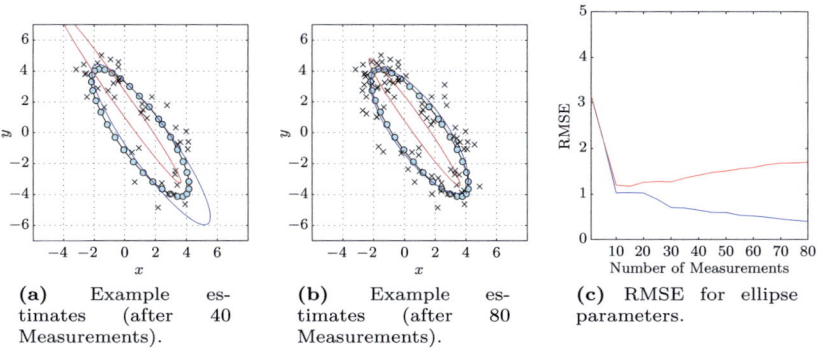

(a) Example es-
timates (after 40
Measurements).

(b) Example es-
timates (after 80
Measurements).

(c) RMSE for ellipse
parameters.

Figure 3.14: Estimation results for Scenario 1b.

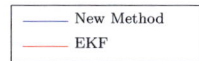

Figure 3.15: Legend for Scenario 1.

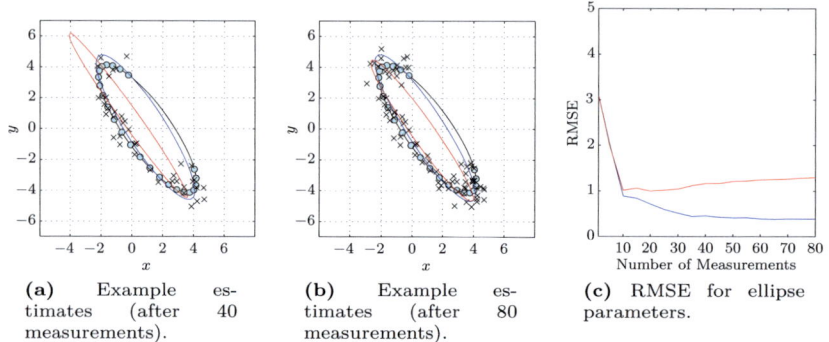

(a) Example estimates (after 40 measurements).

(b) Example estimates (after 80 measurements).

(c) RMSE for ellipse parameters.

Figure 3.16: Estimation results for Scenario 2a.

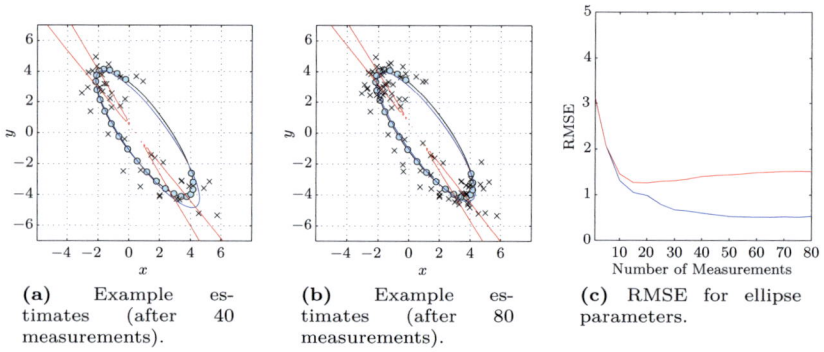

(a) Example estimates (after 40 measurements).

(b) Example estimates (after 80 measurements).

(c) RMSE for ellipse parameters.

Figure 3.17: Estimation results for Scenario 2b.

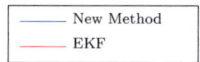

Figure 3.18: Legend for Scenario 2.

In this chapter, we have developed efficient Gaussian filters for conic fitting that perform very well under typical conditions of an extended object tracking scenario. The basic idea is to approximate the original implicit problem with an explicit measurement equation that is corrupted with multiplicative noise. Based on the derived measurement equation, statistical linearization can be performed in order to get a recursive closed-form measurement update. Variants of the proposed filters are used in the following section, where the tracking of region shapes is reduced to the tracking of curves.

Prospective work may concentrate on alternative representations for uncertain ellipses. A probability density for the ellipse parameters has no intuitive meaning (for both discussed parameterizations). Hence, it is difficult to define reasonable system models for the ellipse parameters. Probably a more intuitive representation is possible with the help of random matrix theory [FFK11]. Besides, the proposed techniques may be extended to arbitrary (closed) curves.

A further promising future direction is the investigation of particle filtering techniques based on the derived explicit measurement equation. As conic fitting is a highly nonlinear problem that results in multimodal probability densities, advanced nonlinear estimation techniques such as particles filters probably lead to an increased estimation accuracy for the price of a higher computational complexity.

CHAPTER 4

Random Hypersurface Model

Contents

This chapter is devoted to the tracking of an extended object that is modeled as a region shape, i.e., a closed curve with interior (see Section 1.1). The main contribution of this chapter is a novel extent model called *Random Hypersurface Model* (*RHM*). An *RHM* models the interior of a shape by means of scaling the shape boundary, where the scaling factor is specified by a random variable. In this manner, the problem is reduced to a curve fitting problem and thus, the curve fitting techniques introduced in the previous chapter can be used for estimating region shapes.

An *RHM* is a general concept that is suitable for a wide range of relevant shapes. In this chapter, we introduce specific *RHMs* for estimating the parameters of ellipses and (free-form) star-convex shapes. Based on the *RHMs*, we derive explicit measurement equations and statistical linearization yields efficient Gaussian filters for a closed-form measurement update.

A major highlight of this chapter is the tracking method for star-convex shapes, which is the first with the capability of estimating such detailed shape information. So far only basic shapes such as circle and ellipse have been considered in literature (see also Section 1.6). This is a significant progress in extended object tracking as a detailed shape estimate is an important piece of information, which is valuable for many higher-level information processing tasks such as classification.

The estimation quality and performance of the proposed Gaussian filters are assessed with the help of extensive numerical simulations. We also provide a comparison with a naïve particle filter and a particle filter that

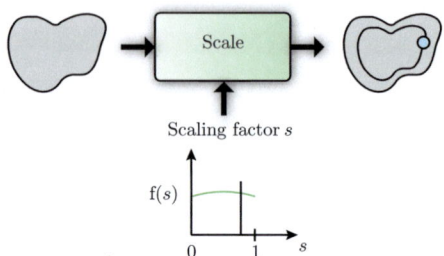

Figure 4.1: Basic concept of a *Random Hypersurface Model.*

employs an approximation of the likelihood function based on *RHMs*. An experimental evaluation of the tracking methods will be given separately in Chapter 6.

Remark 4.1. This chapter is based on the journal publications [1,3] and the conference publications [6,9,14,18,19]. The basic idea was first published in [9]. Elliptic shapes have been treated in [18] and star-convex shapes in [14]. Recent further developments and comparisons of *RHMs* were presented in [6, 19, 22]. This chapter extends previous work on *RHMs* by deriving analytic expressions for the measurement update and a particle filter based on an *RHM*.

4.1 Basic Concept

In this section, the underlying idea of a *Random Hypersurface Models* (*RHMs*) is introduced and a formal definition is given. We focus on *RHMs* for *star-convex region shapes* in N-dimensional space. Nevertheless, *RHMs* are a very general concept, which can be used for many other types of shapes, e.g. surfaces in three-dimensional space.

Definition 4.1 (Star-Convex Shape). A region shape $\mathcal{S} \subset \mathbb{R}^N$ is called *star-convex* (with respect to the origin) iff for all $\underline{z} \in \mathcal{S}$, the line segment from the origin $\underline{0}_N$ to $\underline{z}$ is fully contained in $\mathcal{S}$.

A *Random Hypersurface Model* is an extent model that specifies the location of a *single* measurement source on a star-convex shape (see Section 2.1.2). An *RHM* says that the measurement source lies on a scaled version of the shape boundary (see Fig. 4.1), where the scaling factor is modeled as a one-dimensional random variable. We assume that the probability density of the scaling factor is known and independent of the shape parameters. The scaling factor can be interpreted as a noise term that covers the uncertainty of the measurement source on the object. Where on the boundary the measurement source lies is intentionally left open as there are several alternatives to close this degree of freedom (see Remark 4.4).

According to the above discussion, a *Random Hypersurface Model* is formally defined as follows.

Definition 4.2 (Random Hypersurface Model (RHM)). Suppose we are given

- an N-dimensional extended object with star-convex shape $\mathcal{S}(\underline{p}_k) \subset \mathbb{R}^N$ located at $m_k \in \mathbb{R}^N$, and

- a one-dimensional random variable $s_{k,l} \in [0, 1]$.

Then, the measurement source $\underline{z}_{k,l} \in \mathcal{M}(\underline{p}_k, \underline{m}_k)$ is generated according to an *RHM* with scaling factor $s_{k,l}$ if

$$\underline{z}_{k,l} \in \underline{m}_k + s_{k,l} \cdot \partial \mathcal{S}(\underline{p}_k) \ ,$$

where $\partial \mathcal{S}(\underline{p}_k) \subset \mathbb{R}^N$ denotes the boundary of $\mathcal{S}(\underline{p}_k)$ and the algebraic operations "$\cdot$","+" are interpreted element-wise.

Remark 4.2. It is essential to note that an *RHM* is a probabilistic model, i.e., it involves random quantities. However, an estimator based on an *RHM* may be deterministic (for given measurements).

Remark 4.3. Definition 4.2 is confined to star-convex shapes as they are contractible, i.e., they can be continuously shrunk to a point by scaling. By this means, it is guaranteed that a measurement source lies on the object, i.e., $s_{k,l} \cdot \partial \mathcal{S}(\underline{p}_k) \subset \mathcal{S}(\underline{p}_k)$ for all $s_{k,l} \in [0, 1]$. Mathematically, the scaling of the object boundary corresponds to a straight-line homotopy of the object boundary to the center.

Remark 4.4. Definition 4.2 leaves open where on the scaled boundary the measurement source $\underline{z}_{k,l}$ lies. In general, it can be treated as an unknown parameter (*functional model*) or it can be considered as a random drawn from a potentially unknown probability distribution (*structural model*).

Actually, both models are also used in curve fitting [Che10] and Errors-In-Variables (EIV) models [CRSC06]. For the structural model with a known probability distribution, an *RHM* turns into a *spatial distribution model* [GS05,GGMS05]. An *RHM* can then be seen as a systematic approach for specifying spatial distributions.

4.1.1 Benefits

An *RHM* is a concept for specifying the location of a measurement source on an extended object. The main benefit of an *RHM* is that the modeling of a *region shape* is reduced to the modeling of a *curve* by means of the random scaling factor. For a fixed scaling factor, e.g., $s_{k,l} = 1$, the measurement source lies on the one-dimensional shape boundary, i.e., $\underline{z}_{k,l} \in \partial \mathcal{S}(\underline{p}_k)$, which is a curve. In this sense, an *RHM* comprises curve fitting as a special case. The reduction to curve fitting paves the way for utilizing basic curve fitting techniques. For example, it enables to derive implicit measurement equations as many relevant curves such as ellipses, can be specified by implicit functions. Based upon these measurement equations, statistical linearization allows for an efficient recursive measurement update.

An *RHM* for star-convex shapes as in this chapter can also be interpreted as a reformulation of the likelihood in polar coordinates, i.e., the likelihood function is decomposed into the distance and angle of a measurement source (with respect to the center). This decomposition facilitates the derivation of approximations for the likelihood.

4.1.2 Usage of an RHM

With the help of an *RHM*, it is possible to systematically derive Bayesian estimators for extended objects according to the following steps.

- A shape $\mathcal{S}(\underline{p}_k)$ with a suitable parameterization $\underline{p}_k$ has to be chosen. In general, the shape $\mathcal{S}(\underline{p}_k)$ can be specified as the solution of an implicit equation (see Section 4.2) or explicitly in parametric form (see Section 4.3).

- A probability distribution for the scaling factor $s_{k,l}$ must be determined. Section 4.1.3 discusses how this scaling factor could be chosen.

- Based on the concept of an *RHM*, a measurement equation that relates the parameter vector $\underline{p}_k$, the scaling factor $s_{k,l}$, the measurement noise $\underline{v}_{k,l}$, and the measurement $\hat{\underline{y}}_{k,l}$, has to be derived.

- The measurement equation serves as the basis for constructing a Bayesian state estimator for the shape and the kinematic parameters. In this thesis, we tackle the nonlinearity of the measurement equation by means of statistical linearization as described in see Section 2.4. Nevertheless, *RHMs* make it also possible to derive precise approximations of the likelihood function when a particle filter is desired (see Section 4.3.4).

In this thesis, specific estimators for elliptic shapes (see Section 4.2) and general star-convex shapes (see Section 4.3) are presented following the above steps.

4.1.3 Scaling Factor

The probability distribution of the scaling factor characterizes the locations of the measurement sources on the object. It is object dependent, i.e., it may differ from object to object. For star-convex shapes, there is a reasonable choice for the probability distribution of the scaling factor as illustrated in the following.

If no information about possible measurement sources on an extended object is available, it is reasonable to impose a uniform distribution on the measurement sources. Hence, one could ask the following question: If a measurement source is uniformly distributed on the object, what is the distribution of the scaling factor? Actually, for two-dimensional star-convex shapes, it can be proven that the *squared* scaling factor is uniformly distributed on the interval $[0, 1]$.

Theorem 4.1. *In case the measurement source $\underline{z}$ is uniformly distributed on the two-dimensional star-convex set $\mathcal{S} \subset \mathbb{R}^2$, the squared scaling factor s^2 is uniformly distributed on the interval $[0, 1] \subset \mathbb{R}$.*

Proof. The cumulative distribution function of $\boldsymbol{s}$ is given by

$$
\begin{aligned}
F(s) &= \mathrm{P}\left(\boldsymbol{s} \le s\right) \\
&= \mathrm{P}\left(\underline{\boldsymbol{z}} \in s \cdot \mathcal{S}\right) = \frac{\text{Area}(s \cdot \mathcal{S})}{\text{Area}(\mathcal{S})} \\
&= s^2
\end{aligned}
$$

for $s \in [0, 1]$. Furthermore, $F(s) = 0$ for $s < 0$ and $F(s) = 1$ for $s > 1$. Hence, the probability density function of $\boldsymbol{s}$ turns out to be

$$
f_s(s) = \begin{cases} 2s & s \in [0, 1] \\ 0 & \text{otherwise} \end{cases} \quad . \tag{4.1}
$$

In general, if $\boldsymbol{s} \sim f_s(s)$, then the density of $\boldsymbol{u} := \boldsymbol{s}^2$ is given by [PP02, p. 125]

$$
f_u(u) = \begin{cases} \frac{1}{2\sqrt{u}} \cdot \left(f_s(\sqrt{u}) + f_s(-\sqrt{u})\right) & u > 0 \\ 0 & \text{otherwise} \end{cases} \quad . \tag{4.2}
$$

Substituting (4.1) in (4.2) gives

$$
f_u(u) = \begin{cases} 1 & u \in [0, 1] \\ 0 & \text{otherwise} \end{cases} \quad .
$$

$\square$

Theorem 4.1 justifies to impose a uniform distribution on the squared scaling factor. The converse of Theorem 4.1 is not true, i.e., a uniformly distributed squared scaling factor is *not* necessarily the result of a uniformly distributed measurement source on the object. There is more than one spatial distribution that gives rise to a uniformly distributed squared scaling factor.

Remark 4.5. It may be suitable to approximate the probability distribution of the scaling factor with a Gaussian distribution, e.g., the mean $\frac{1}{2}$ and variance $\frac{1}{12}$ resulting from moment matching with a uniform distribution are reasonable.

4.2 Elliptic Shapes

In this section, we employ an *RHM* as introduced in the previous section for tracking an elliptic region. Elliptic shapes are relevant for a variety of applications as the shape of many real-world objects is approximately elliptic. The orientation and length of the semi-axes provide valuable information that can be extracted even under rather difficult conditions, e.g., high measurement noise and few available measurements.

4.2.1 Implicit Representation of an Ellipse

Based on the insights gained from conic fitting, we use the center/shape representation for an ellipse (see Section 3.5), i.e., the state vector is given by $\underline{x}_k^{\mathrm{el}} = \left[\underline{m}_k^T, (\underline{x}_k^*)^T, (\underline{p}_k^{\mathrm{el}})^T \right]^T$, where $\underline{m}_k$ is the center, $\underline{x}_k^*$ encompasses variables for the kinematics, and $\underline{p}_k^{\mathrm{el}} := \left[l_k^{(1)}, l_k^{(2)}, l_k^{(3)} \right]^T$ determines the (two-dimensional) ellipse.

The ellipse specified by $\underline{x}_k^{\mathrm{el}}$ is given by all points $\underline{z} \in \mathbb{R}^2$ that satisfy the implicit quadratic function

$$(\underline{z} - \underline{m}_k)^T \mathbf{B}_k^{-1} (\underline{z} - \underline{m}_k) - 1 = 0 \; ,$$

where $\mathbf{B}_k^{-1} = \mathbf{L}_k \mathbf{L}_k^T$ with

$$\mathbf{L}_k := \begin{bmatrix} l_k^{(1)} & 0 \\ l_k^{(3)} & l_k^{(2)} \end{bmatrix} \; . \tag{4.3}$$

The corresponding shape is

$$\mathcal{S}^{\mathrm{el}}(\underline{p}_k^{\mathrm{el}}) = \{ \underline{z} \in \mathbb{R}^2 \mid \underline{z}^T \mathbf{B}_k^{-1} \underline{z} - 1 \le 0 \} \tag{4.4}$$

and the measurement sources are given by

$$\mathcal{M}^{\mathrm{el}}(\underline{p}_k^{\mathrm{el}}, \underline{m}_k) := \mathcal{S}^{\mathrm{el}}(\underline{p}_k^{\mathrm{el}}) + \underline{m}_k \; . \tag{4.5}$$

4.2.2 Measurement Equation

In the following, a measurement equation is derived based on the idea of an *RHM*. Due to the chosen center/shape representation of an ellipse, the scaled version of the ellipse boundary $\partial \mathcal{S}^{\mathrm{el}}(\underline{p}_k^{\mathrm{el}})$ with scaling factor s is given by

$$\begin{aligned} s \cdot \partial \mathcal{S}^{\mathrm{el}}(\underline{p}_k^{\mathrm{el}}) &= s \cdot \{ \underline{z} \in \mathbb{R}^2 \mid \underline{z}^T \mathbf{B}_k^{-1} \underline{z} - 1 = 0 \} \\ &= \{ \underline{z} \in \mathbb{R}^2 \mid \underline{z}^T \mathbf{B}_k^{-1} \underline{z} - s^2 = 0 \} \; . \end{aligned}$$

Hence, a measurement source $\underline{z}_{k,l}$ can be related to the ellipse state $\underline{x}_k^{\mathrm{el}}$ and the scaling factor $s_{k,l}^2$ by

$$g^{\mathrm{el}}(\underline{x}_k^{\mathrm{el}}, s_{k,l}, \underline{z}_{k,l}) = 0 \ , \tag{4.6}$$

where

$$g^{\mathrm{el}}(\underline{x}_k^{\mathrm{el}}, s_{k,l}, \underline{z}_{k,l}) := (\underline{z}_{k,l} - \underline{m}_k)^T \cdot \mathbf{B}_k^{-1} \cdot (\underline{z}_{k,l} - \underline{m}_k) - s_{k,l}^2 \ . \tag{4.7}$$

The random variable $s_{k,l}^2$ can be interpreted as a noise term modeling the uncertainty arising from the unknown measurement source.

When there is no measurement noise, i.e., $\underline{z}_{k,l} = \hat{\underline{y}}_{k,l}$, (4.6) can directly be seen as the measurement equation. Moreover, usually only a noisy measurement $\hat{\underline{y}}_{k,l}$ of the measurement source $\underline{z}_{k,l}$ is given according to measurement equation (2.2)

$$\hat{\underline{y}}_{k,l} = \underline{z}_{k,l} + \underline{v}_{k,l} \ . \tag{4.8}$$

The estimation problem specified by (4.6) and (4.8) nearly coincides with the curve fitting problem treated in Chapter 3. The only difference is the additional scaling factor $s_{k,l}^2$, which can be interpreted as an additional noise term. For this reason, we can use the curve fitting techniques from Chapter 3 with a slight modification that incorporates $s_{k,l}^2$. The first step is to put the measurement $\hat{\underline{y}}_{k,l}$ in (4.6), which yields after some algebraic manipulations

$$
\begin{aligned}
g^{\mathrm{el}}(\underline{x}_k^{\mathrm{el}}, s_{k,l}, \hat{\underline{y}}_{k,l}) &= g^{\mathrm{el}}(\underline{x}_k^{\mathrm{el}}, s_{k,l}, \underline{z}_{k,l} + \underline{v}_{k,l}) \\
&= \underbrace{g^{\mathrm{el}}(\underline{x}_k^{\mathrm{el}}, s_{k,l}, \underline{z}_{k,l})}_{=0} + 2(\underline{z}_{k,l} - \underline{m}_k)^T \mathbf{B}_k^{-1} \underline{v}_{k,l} \\
&\quad + \underline{v}_{k,l}^T \mathbf{B}_k^{-1} \underline{v}_{k,l} \ ,
\end{aligned}
$$

where $\mathbf{B}_k$ is the shape matrix specified by $\underline{p}_k^{\mathrm{el}}$. Hence, the following measurement equation $h^{\mathrm{el}}(\underline{x}_k^{\mathrm{el}}, \underline{v}_{k,l}, s_{k,l}, \hat{\underline{y}}_{k,l})$ is obtained

$$
\begin{aligned}
0 &= g^{\mathrm{el}}(\underline{x}_k^{\mathrm{el}}, s_{k,l}, \hat{\underline{y}}_{k,l}) - 2(\underline{z}_{k,l} - \underline{m}_k)^T \mathbf{B}_k^{-1} \underline{v}_{k,l} - \underline{v}_{k,l}^T \mathbf{B}_k^{-1} \underline{v}_{k,l} \\
&=: h^{\mathrm{el}}(\underline{x}_k^{\mathrm{el}}, \underline{v}_{k,l}, s_{k,l}, \hat{\underline{y}}_{k,l}) \ , \tag{4.9}
\end{aligned}
$$

where $h^{\mathrm{el}}(\underline{x}_k^{\mathrm{el}}, \underline{v}_{k,l}, s_{k,l}, \hat{\underline{y}}_{k,l})$ maps the state $\underline{x}_k^{\mathrm{el}}$, the measurement noise $\underline{v}_{k,l}$, the scaling factor $s_{k,l}$, and the measurement $\hat{\underline{y}}_{k,l}$ to the pseudo-measurement 0.

Note that the unknown measurement source $\underline{z}_{k,l}$ in (4.9) can be substituted by a point estimate. For example, a suitable point estimate for $\underline{z}_{k,l}$ is given by the point with the smallest distance from the current ellipse estimate, i.e., $\underline{\mu}_{k,l-1}^p$, to the measurement $\hat{\underline{y}}_{k,l}$.

Remark 4.6. Of course, it would be also possible to substitute $\hat{\underline{y}}_{k,l} - \underline{v}_{k,l}$ for the measurement source in (4.9). However, in the same way as for circle fitting in Section 3.4.3, we observed that a Gaussian filter based on statistical linearization for this equation gives biased estimates for large measurement noise.

4.2.3 Gaussian Filter Using Statistical Linearization

The measurement equation (4.9) is quadratic and corrupted with multiplicative Gaussian noise. In the same way as in Section 2.4 and Section 3.3, the nonlinearity can be tackled with statistical linearization. The density $f_{l-1}(\underline{x}_k) = \mathcal{N}(\underline{x}_k - \underline{\mu}_{k,l-1}^x, \Sigma_{k,l-1}^x)$ is updated with the measurement $\hat{\underline{y}}_{k,l}$ according to

$$
\underline{\mu}_{k,l}^x = \underline{\mu}_{k,l-1}^x + \Sigma_{k,l}^{xh} \left(\Sigma_{k,l}^{hh}\right)^{-1} \left(0 - \mu_{k,l}^h\right) \; ,
$$
$$
\Sigma_{k,l}^x = \Sigma_{k,l-1}^x - \Sigma_{k,l}^{xh} \left(\Sigma_{k,l}^{hh}\right)^{-1} \Sigma_{k,l}^{hx} \; ,
$$

where $\mu_{k,l}^h$ is the predicted pseudo-measurement, $\Sigma_{k,l}^{xh}$ is the covariance between the pseudo-measurement and state, and $\Sigma_{k,l}^{hh}$ is the variance of the predicted pseudo-measurement.

As discussed in Section 2.4, the predicted measurement $\mu_{k,l}^h$, the covariance matrices Σ_k^{xh}, and $\Sigma_{k,l}^{hh}$ can be calculated approximately with the unscented transform [JU04] or even analytically with moment calculation as described in Section 2.4.1. As both $\mu_{k,l}^h$ and Σ_k^{xh} are independent of the unknown measurement source $\underline{z}_{k,l}$, the error made due to the point estimate is rather negligible. In the remainder, the Gaussian filter based on the unscented transform is referred to as *UKF-EL-RHM*.

4.2.4 Evaluation

In the following, the proposed tracking algorithm *UKF-EL-RHM* for elliptic shapes is evaluated in two scenarios. In the first scenario, a stationary,
i.e, non-moving, extended object is considered. The purpose of the first
scenario is to show the robustness of the method against errors regarding
the true shape and to illustrate how the precision of the shape estimate
depends on the measurement noise level. The second scenario shows the
practicability of the *UKF-EL-RHM* for tracking an extended object whose
dynamics is modeled as a constant velocity model.

Stationary Extended Object

In the first scenario, 200 measurements are received sequentially, i.e., a
single measurement per time step is available ($n_k = 1$), from a single
extended object with *fixed* position and shape. Simulations are carried
out with

- low measurement noise level $\Sigma_{k,1}^v = \mathrm{diag}(0.6, 0.6)$,

- medium measurement noise level $\Sigma_{k,1}^v = \mathrm{diag}(1, 1)$, and

- high measurement noise level $\Sigma_{k,1}^v = \mathrm{diag}(1.4, 1.4)$.

The two different object types as illustrated in Fig. 4.2 are considered.
The first object is an elliptic region in which the measurement sources are
uniformly distributed and the second object is a group object from which
the measured point objects are selected uniformly.

The shape is estimated with an *RHM* for ellipses coupled with the *UKF*,
i.e., the *UKF-EL-RHM*, where the measurements are incorporated sequentially. The squared scaling factor is Gaussian distributed with mean 0.5
and variance 0.06. Furthermore, the state vector $\underline{x}_k^{\mathrm{el}}$ is initialized with
a Gaussian with mean $\underline{\mu}_{0,0}^x = [0.5, 0.5, 1.6, 1.6, 0]^T$ and covariance matrix
$\Sigma_{0,0}^x = \mathrm{diag}(3, 3, 0.5, 0.5, 0.5)$, i.e., an uncertain circle with radius 1.2 and
center $[0.5, 0.5]^T$.

The estimation results for the two objects are shown in Fig. 4.4 and Fig. 4.6
(after 50 measurements and after all 200 measurements). The corresponding shape estimates, i.e., the shape parameters, are averaged over 20 Monte

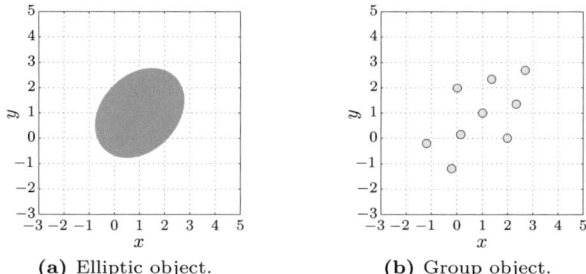

(a) Elliptic object.   (b) Group object.

Figure 4.2: Extended objects used for evaluating *UKF-EL-RHM*.

Carlo runs. In order to convey an impression of the magnitude of the measurement noise, the example measurements of a particular run are depicted in Fig. 4.3 and Fig. 4.5.

The results show that the shapes are estimated precisely for both kinds of extended objects. As the first object (see Fig. 4.2a) is an elliptic region from which the measurement sources are drawn uniformly, an *RHM* for ellipses with a uniformly distributed squared scaling factor (see Theorem 4.1) correctly models the shape, i.e., there is no modeling mismatch. With an increasing magnitude, the measurement noise dominates the extent and the orientation of the ellipse estimate gets slightly lost. For the group object in Fig. 4.2b, the corresponding (true) squared scaling factor is in fact not uniformly distributed. Nevertheless, even for this group object, the ellipse approximation according to an *RHM* is accurate.

In summary, the above examples demonstrate the *UKF-EL-RHM* implementation of an *RHM* gives precise shape estimates even in case of extremely high measurement noise. Furthermore, the *UKF-EL-RHM* is robust against different object types as good shape estimates are also obtained for group objects.

Tracking an Elliptic Shape

In the second scenario, an aircraft-shaped object (see Fig. 4.7a) is tracked using the *UKF-EL-RHM* and a (nearly) constant velocity model for the object's motion (see Section 2.1.3). The object follows the trajectory depicted in Fig. 4.7b.

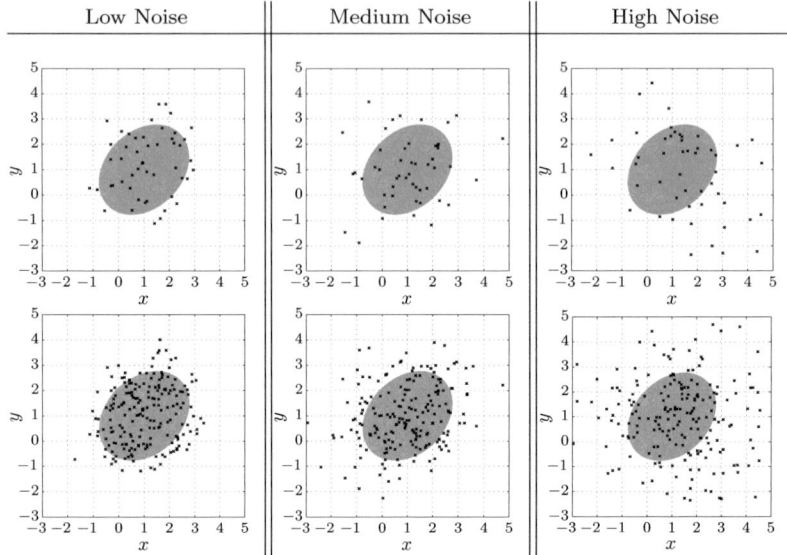

Figure 4.3: Object 1: 200 example measurements (second row) and the first 50 measurements (first row).

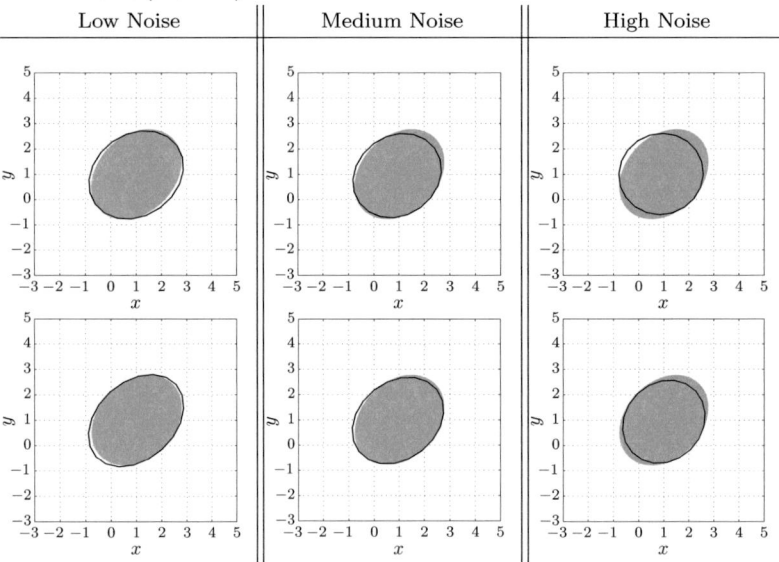

Figure 4.4: Object 1: Shape estimates averaged over 20 runs after 50 and 200 measurements.

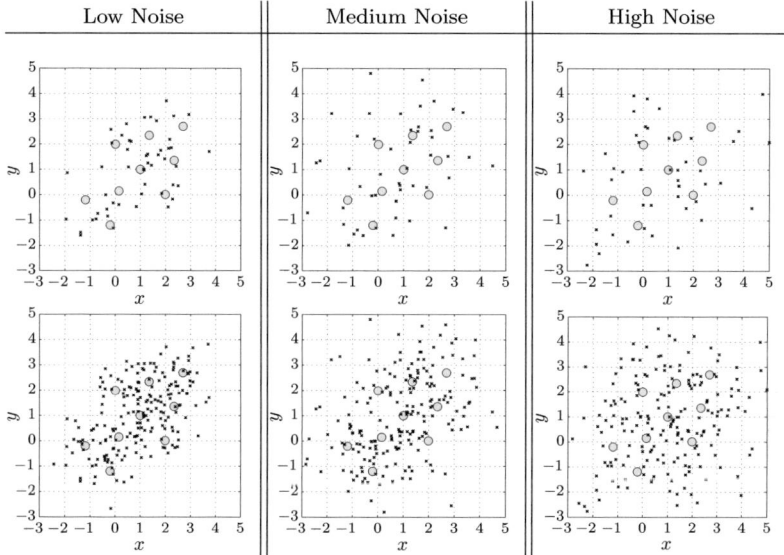

Figure 4.5: Object 2: 200 example measurements (second row) and the first 50 measurements (first row).

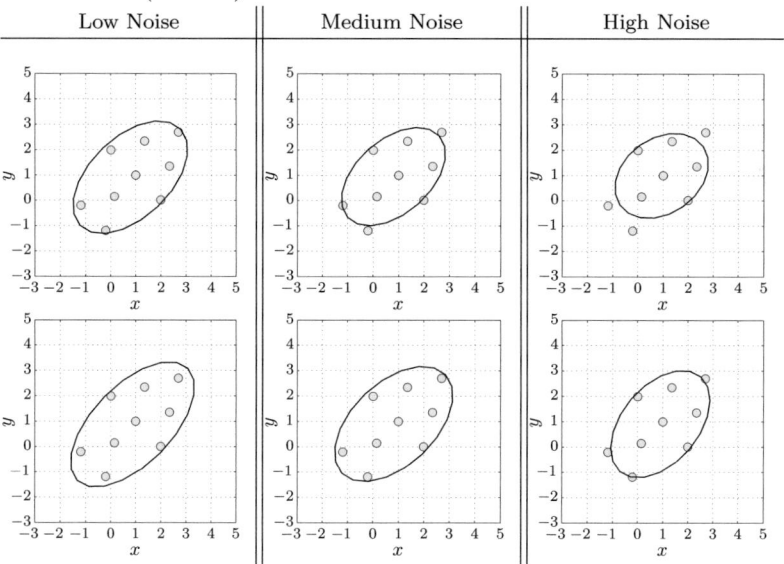

Figure 4.6: Object 2: Shape estimates averaged over 20 runs after 50 and 200 measurements.

Measurement sources are drawn uniformly from the object's surface and the number of measurements received per time instant is $n_k = n_k^* + 1$, where n_k^* is Poisson distributed with mean 4. Hence, it is even possible that only a single measurement per time step is given. The magnitude of the measurement noise differs from measurement to measurement in order to emulate different sensors, i.e., the covariance matrix of the measurement noise is $\Sigma_{k,1}^v = \mathrm{diag}(0.2, 0.2)$ with probability 0.6 and $\Sigma_{k,1}^v = \mathrm{diag}(0.8, 0.8)$ with probability 0.4.

The state vector to be tracked is $\underline{x}_k^{\mathrm{el}} = \left[\underline{m}_k^T, (\underline{m}_k^v)^T, (\underline{p}_k^{\mathrm{el}})^T \right]^T$, where $\underline{m}_k$ is the center, $\underline{m}_k^v$ is the velocity vector, and $\underline{p}_k^{\mathrm{el}}$ are the ellipse parameters. We use the motion model (2.5), where the center is modeled as a constant velocity model and a random walk model is used for the shape parameters. The noise parameters for (2.5) are $\mathbf{C}_k^p = 0.0015 \cdot \mathbf{I}_5$ and $\mathbf{C}_k^{cv} = 0.005 \cdot \mathbf{I}_2$. The squared scaling factor is Gaussian distributed with mean 0.5 and variance 0.06. The shape estimates are depicted in Fig. 4.7 for two details of the trajectory (averaged over 20 time steps). The first detail in Fig. 4.7d shows that the shape is precisely estimated, i.e., the orientation and size fits well. As soon as the object enters the curve, the orientation changes are followed with a slight delay due to the high measurement noise and the very few available measurements (see Fig. 4.7f).

Comparison with the Random Matrix Approach   A detailed comparison of elliptic *RHMs* and the random matrix approach, which is also capable of tracking ellipsoidal shapes, is given in [6]. The main difference is the description of an uncertain ellipse, i.e., in contrast to random matrices, *RHMs* work with Gaussian distributions. Both approaches have to perform approximations in the measurement update step.

4.3   Star-Convex Shapes

Previous extended object tracking methods for region shapes only consider basic shapes such as ellipses or circles (see Section 1.6). In this section, the next step is performed by introducing an extended object tracking method that is capable of estimating the parameters of a *free-form star-convex shape*. This is a substantial progress in extend object tracking as the shape is very important and relevant information.

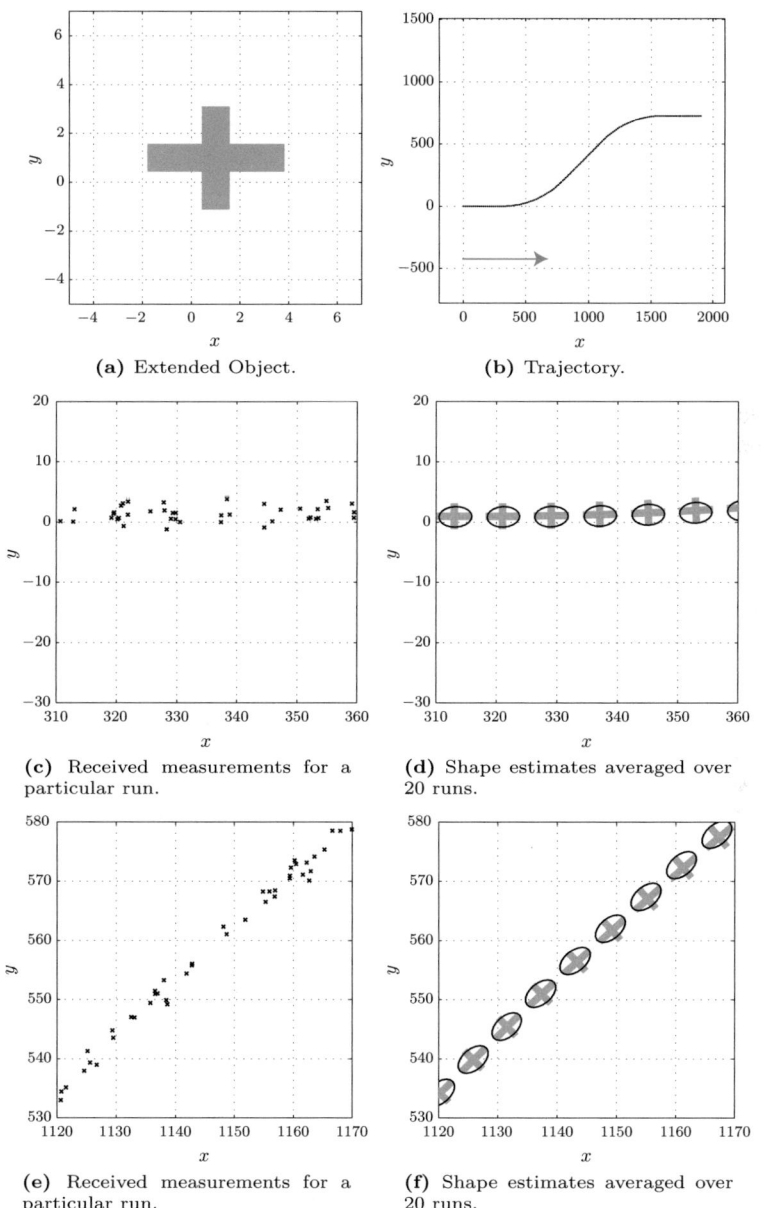

(a) Extended Object.

(b) Trajectory.

(c) Received measurements for a particular run.

(d) Shape estimates averaged over 20 runs.

(e) Received measurements for a particular run.

(f) Shape estimates averaged over 20 runs.

Figure 4.7: Tracking an extended object with an *RHM* for elliptic objects and a constant velocity model.

First, it is obvious that a more detailed shape model improves the estimation quality for the kinematic state as it better captures the real object. Furthermore, the shape of the object is inherently coupled with the object's motion. For example, the object may always be aligned with its motion direction. Both the estimates for the shape and kinematic parameters can benefit from this coupling.

Second, many other (higher level) information processing algorithms exploit the shape estimate as described in the following.

- *Target classification*
  In many applications such as surveillance, the type of the object is totally unknown when the track is initialized. A detailed shape estimate is a valuable source of information that can facilitate the classification of the object. For example, in air surveillance the shape may allow for determining the type of an airplane.

- *Track management*
  Obviously, a detailed shape estimate allows for an early detection of group splitting and merging events. For example, consider a group tracking scenario in which a group member leaves the group. When modeling the shape of the group with an ellipse, the splitting results in an increasing overall size of the group. However, when having a star-convex shape approximation, the splitting appears as a buckle that can be detected very early. Even the motion direction of the leaving group member can be determined.

- *Data association*
  Typically, one has to deal with multiple objects and false measurements, i.e., measurements not stemming from any object. When dealing with such data association problems, a crucial task is to determine the probability that a particular measurements stems from an object. Of course, the more precise the shape of an object is known, the more precise this probability can be determined.

- *Sensor management and planning*
  Most sensors involve parameters to adjust, e.g., the viewing direction. Typically, a sensor management algorithm tweaks these parameters so that future measurements provide as much as information as possible. When the object is extended, a sensor management algorithm

may incorporate the extent in order to improve the information gain. Furthermore, it is obvious that a detailed shape estimate is essential when performing motion planning. For example, a mobile robot may have to navigate through a group of people.

Of course, a detailed shape estimate can only be extracted under rather good conditions, i.e.,

- the measurement noise is low in comparison to the object's extent,

- a sufficient number of measurements is available, or

- the object evolves rather slowly, e.g., no fast maneuver is performed.

If this is not the case, a rather coarse shape such as an ellipse should be preferred (see also the discussion in Section 1.1).

The remainder of this section is about estimating a star-convex shape approximation of an extended object with the help of an *RHM*. For this purpose, a star-convex shape is modeled with a polar function, which specifies the distance from the center to a boundary point for a given angle. The shape parameters that are to be estimated are given by the first Fourier coefficients of the Fourier expansion of the polar function. Based on this shape representation and the concept of an *RHM*, we will derive three Bayesian estimators, i.e.,

- a Gaussian filter using the *UKF* called *UKF-SC-RHM*,

- a Gaussian filter using analytic moment calculation *AMC-SC-RHM*,

- a simple particle filter *PF-SC-RHM* that uses the concept of an *RHM* for approximating the likelihood function.

The performance of the presented filters is assessed in Section 4.3.5. It turns out that the Gaussian filters significantly outperform the particle filter *PF-SC-RHM* and also the naïve particle filter (see Section 2.3) in both computational complexity and accuracy.

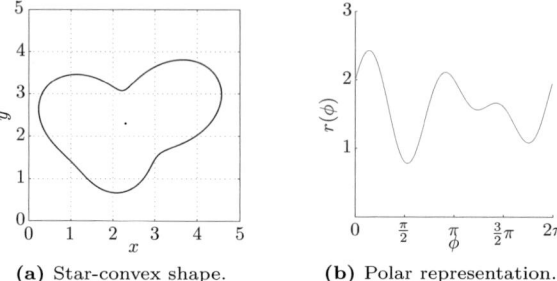

(a) Star-convex shape. (b) Polar representation.

Figure 4.8: Representing a star-convex shape with a polar function.

4.3.1 Parametric Representation of Star-Convex Shapes

A star-convex shape is represented in *parametric form* with a polar function (see [ZL05, JKS95, DN99]), which specifies the distance $r(\phi)$ from the center to a boundary point depending on the angle ϕ (see Fig. 4.8). The region enclosed by the boundary specified by a polar function is always star-convex.

In order to specify the polar function with a finite dimensional parameter vector $\underline{p}_k^{\mathrm{sc}}$, the first N_F Fourier coefficients of the Fourier expansion [ZL05] of $r(\phi)$ are employed, i.e.,

$$r(\underline{p}_k^{\mathrm{sc}}, \phi) := \frac{\boldsymbol{a}_k^{(0)}}{2} + \sum_{j=1\ldots N_F} \boldsymbol{a}_k^{(j)} \cos(j\phi) + \boldsymbol{b}_k^{(j)} \sin(j\phi) \ , \qquad (4.10)$$

where $\underline{p}_k^{\mathrm{sc}}$ denotes the parameter vector given by

$$\underline{p}_k^{\mathrm{sc}} := \left[\boldsymbol{a}_k^{(0)}, \boldsymbol{a}_k^{(1)}, \boldsymbol{b}_k^{(1)}, \ldots \boldsymbol{a}_k^{(N_F)}, \boldsymbol{b}_k^{(N_F)} \right]^T \ .$$

For fixed ϕ, (4.10) is linear in $\underline{p}_k^{\mathrm{sc}}$, i.e.,

$$r(\underline{p}_k^{\mathrm{sc}}, \phi) = \underline{q}(\phi) \cdot \underline{p}_k^{\mathrm{sc}} \ ,$$

where

$$\underline{q}(\phi) := \left[1, \cos(\phi), \sin(\phi), \ldots, \cos(N_F\phi), \sin(N_F\phi) \right] \ . \qquad (4.11)$$

The Fourier coefficients have an intuitive meaning: Fourier coefficients
with lower indices capture coarse features of the shape and Fourier coefficients with higher indices encode finer details. Note that other shape
representations, e.g., splines [BI98], may also be suitable for representing
star-convex shapes.

In summary, the overall state vector for a star-convex extended object is
given by $\underline{x}_k^{sc} = \left[\underline{m}_k^T, (\underline{x}_k^*)^T, (\underline{p}_k^{sc})^T \right]^T$, which contains the center $\underline{m}_k$ of the
object, further optional variables $\underline{x}_k^*$, and the Fourier coefficients $\underline{p}_k^{sc}$.

4.3.2 Measurement Equation

Following the idea of an *RHM*, the region enclosed by the star-convex
contour specified by $\underline{p}_k^{sc}$ can be written as

$$\mathcal{S}^{sc}(\underline{p}_k^{sc}) := \left\{ s \cdot r(\underline{p}_k^{sc}, \phi_{k,l}) \cdot \underline{e}(\phi_{k,l}) \mid \phi_{k,l} \in [0, 2\pi] \text{ and } s \in [0,1] \right\} \ ,$$
$$(4.12)$$

where $\underline{e}(\phi_{k,l}) := [\cos(\phi_{k,l}), \sin(\phi_{k,l})]^T$ is a unit vector with angle $\phi_{k,l}$ to
the x-axis and s is a scaling factor that captures the interior of the shape.
The set of measurement sources forming the extended object is given by
(see (2.1))

$$\mathcal{M}^{sc}(\underline{p}_k^{sc}, \underline{m}_k) := \mathcal{S}^{sc}(\underline{p}_k^{sc}) + \underline{m}_k \ , \qquad (4.13)$$

which results from shifting the shape to the location of the extended
object $\underline{m}_k$.

Based on (2.2) and (4.13), the following explicit measurement equation is
obtained

$$
\begin{aligned}
\hat{\underline{y}}_{k,l} &= \underline{z}_{k,l} + \underline{v}_{k,l} \\
&= s_{k,l} \cdot \underline{q}(\phi_{k,l}) \cdot \underline{p}_k^{sc} \cdot \underline{e}(\phi_{k,l}) + \underline{m}_k + \underline{v}_{k,l} \qquad (4.14) \\
&=: h_{\text{expl}}^{sc}(\underline{x}_k^{sc}, \underline{v}_{k,l}, s_{k,l}, \phi_{k,l}) \ ,
\end{aligned}
$$

where $h_{\text{expl}}^{sc}(\underline{x}_k^{sc}, \underline{v}_{k,l}, s_{k,l}, \phi_{k,l})$ maps the state $\underline{x}_k^{sc}$, the measurement noise
$\underline{v}_{k,l}$, the scaling factor $s_{k,l}$, and the angle $\phi_{k,l}$ to the measurement $\hat{\underline{y}}_{k,l}$.

A reasonable assumption is to choose a probability distribution for $\phi_{k,l}$
such that the resulting measurement sources are uniformly distributed
over the shape boundary. However, this probability distribution depends

on the actual shape, i.e., $\underline{p}_k^{\mathrm{sc}}$, and no general expressions can be derived. In section Section 4.3.4, we derive an approximation for the likelihood function based on (4.14) for the use within particle filters.

Unfortunately, (4.14) is rather unsuitable for statistical linearization aiming at a Gaussian filter; already the polar equation for circle fitting in Section 3.4.1, which is a special case of (4.14), yields unsatisfactory results.

In order to perform statistical linearization the following reformulation of (4.14) is performed:

$$\underline{\hat{y}}_{k,l} - \underline{m}_k \;=\; s_{k,l} \cdot \underline{e}(\phi_{k,l}) \cdot \underline{q}(\phi_{k,l}) \cdot \underline{p}_k^{\mathrm{sc}} + \underline{v}_{k,l} \;,$$

and

$$||\underline{\hat{y}}_{k,l} - \underline{m}_k||^2 \;=\; s_{k,l}^2 \cdot ||\underline{q}(\phi_{k,l}) \cdot \underline{p}_k^{\mathrm{sc}}||^2 + 2 \cdot s_{k,l} \cdot \underline{q}(\phi_{k,l}) \cdot \underline{p}_k^{\mathrm{sc}}$$
$$\cdot \underline{e}(\phi_{k,l})^T \cdot \underline{v}_{k,l} + ||\underline{v}_{k,l}||^2 \;. \tag{4.15}$$

The above reformulation is inspired by the findings from circle fitting. Essentially, it corresponds to a "circulization" of the shape, i.e., the shape is assumed to be approximately circular around a particular ϕ. Note the similarity to the measurement equation (3.21) for circle fitting.

Taking the squared norm can be interpreted as a projection on the (squared) distance. By this means, the direction vector $\underline{e}(\phi_{k,l})$ in the first summand of (4.15) is removed. The second summand can be interpreted as a noise term, which depends on the state. The remaining unknown values of $\phi_{k,l}$ in (4.15), can be substituted with a point estimate. A proper simple point estimate $\phi_{k,l}$ is the angle enclosed by the vector from the current shape center estimate $\underline{\mu}_{k,l-1}^m$ to the measurement $\underline{\hat{y}}_{k,l}$ and the x-axis

$$\hat{\phi}_{k,l} := \angle\left(\underline{\hat{y}}_{k,l} - \underline{\mu}_{k,l-1}^m, [1,0]^T\right) \;.$$

Based on (4.15) and the point estimate, the following measurement equation can be defined as

$$0 \;=\; s_{k,l}^2 \cdot ||\underline{q}(\hat{\phi}_{k,l}) \cdot \underline{p}_k^{\mathrm{sc}}||^2 + 2 \cdot s_{k,l} \cdot \underline{q}(\hat{\phi}_{k,l}) \cdot \underline{p}_k^{\mathrm{sc}} \cdot \underline{e}(\hat{\phi}_{k,l})^T \cdot \underline{v}_{k,l}$$
$$+ ||\underline{v}_{k,l}||^2 - ||\underline{\hat{y}}_{k,l} - \underline{m}_k||^2$$
$$=: \; h^{\mathrm{sc}}(\underline{x}_k^{\mathrm{sc}}, \underline{v}_{k,l}, s_{k,l}, \underline{\hat{y}}_{k,l}) \;, \tag{4.16}$$

which maps the state $\underline{x}_k^{\mathrm{sc}}$, the measurement noise $\underline{v}_{k,l}$, the scaling factor $s_{k,l}$, and the measurement $\underline{\hat{y}}_{k,l}$ to a pseudo-measurement 0.

4.3.3   Gaussian Filter using Statistical Linearization

The quadratic measurement equation (4.16) can be statistically linearized as described in Section 2.4 in order to perform the measurement update with a Gaussian filter. In this manner, the update of the density $f_{l-1}(\underline{x}_k) = \mathcal{N}(\underline{x}_k - \underline{\mu}_{k,l-1}^x, \Sigma_{k,l-1}^x)$ with measurement $\underline{\hat{y}}_{k,l}$ results in $f_l(\underline{x}_k) = \mathcal{N}(\underline{x}_k - \underline{\mu}_{k,l}^x, \Sigma_{k,l}^x)$, where

$$
\begin{aligned}
\underline{\mu}_{k,l}^x &= \underline{\mu}_{k,l-1}^x + \Sigma_k^{xh}(\Sigma_k^{hh})^{-1}\left(0 - \mu_{k,l}^h\right) \;, &(4.17)\\
\Sigma_{k,l}^x &= \Sigma_{k,l-1}^x - \Sigma_k^{xh}(\Sigma_k^{hh})^{-1}\Sigma_k^{hx} \;,
\end{aligned}
$$

with the predicted pseudo-measurement $\mu_{k,l}^h$, covariance $\Sigma_{k,l}^{xh}$ between the pseudo-measurement and current state estimate, and the variance of the pseudo-measurement $\Sigma_{k,l}^{hh}$. Below, we present two methods for calculating the unknown moments in (4.17).

Analytic Expressions (AMC-SC-RHM)

The general procedure described in Section 2.4 allows for deriving analytic expressions for the moments in (4.17). In fact, this procedure expands all polynomials in order to reduce the problem to calculating the expectations of monomials. It is completely automatic, however, the resulting expressions are in general rather complicated as they are not simplified. In the following, we present more compact analytic expressions with the help of well known identities for the expectations of quadratic, cubic, and quartic expressions of Gaussian random variables (see [PP08, Bro11]). As the calculation of these moments is essentially straight-forward, we only state the final result.

We make the simplifying assumption that the center of the object and the shape parameters are uncorrelated and consider a slight approximation of the original measurement equation (4.16). The only reason for this assumption is that it simplifies the resulting expressions; however, it does not necessarily have to be made.

All told, given are (the time index k and measurement number index l are omitted)

- the current state estimate characterized by the mean and covariance matrix of a Gaussian in the form

$$\underline{\mu}^x = \begin{bmatrix} \underline{\mu}^m \\ \underline{\mu}^p \end{bmatrix} \text{ and } \Sigma^x = \begin{bmatrix} \Sigma^m & \mathbf{0} \\ \mathbf{0}^T & \Sigma^p \end{bmatrix} \quad ,$$

- the mean and μ^s covariance matrix Σ^s of the squared scaling factor,

- the covariance matrix of the (zero-mean) measurement noise Σ^v, and

- the following measurement equation

$$\boldsymbol{h} := \underbrace{s^2 \cdot ||\underline{q}(\hat{\phi}) \cdot \underline{p}||^2}_{:=\boldsymbol{h}_1} - \underbrace{||\hat{\underline{y}} - \underline{m}||^2}_{:=\boldsymbol{h}_2} + \underbrace{2\mu^s \underline{q}(\hat{\phi}) \cdot \underline{\mu}^p \cdot \underline{e}(\hat{\phi})^T \cdot \underline{v} + ||\underline{v}||^2}_{:=\boldsymbol{h}_3},$$

which is a slight approximation of measurement equation (4.16). The current estimates are substituted in order to obtain uncorrelated noise.

Desired are the

- mean $\underline{\mu}^h$ and covariance Σ^h of $\boldsymbol{h}$, and the

- cross-covariance matrix $\Sigma^{hx} = \begin{bmatrix} \Sigma^{hm} \\ \Sigma^{hp} \end{bmatrix}$ between $\boldsymbol{h}$ and the state $\underline{x}$.

As the random variables $\boldsymbol{h}_1$, $\boldsymbol{h}_2$, and $\boldsymbol{h}_3$ are stochastically independent, we can write

$$\begin{aligned} \underline{\mu}^h &= \underline{\mu}^{h_1} - \underline{\mu}^{h_2} + \underline{\mu}^{h_3} \quad , \\ \Sigma^h &= \Sigma^{h_1} + \Sigma^{h_2} + \Sigma^{h_3} \quad , \\ \Sigma^{hx} &= \Sigma^{h_1 x} - \Sigma^{h_2 x} + \Sigma^{h_3 x} \quad . \end{aligned}$$

The moments involving $\boldsymbol{h}_1$ are given by

$$
\begin{aligned}
\underline{\mu}^{h_1} &= \mu^{s^2} \cdot \left(\mathrm{trace}\{\mathbf{Q}\Sigma^p + (\underline{\mu}^p)^T \mathbf{Q}\underline{\mu}^p\} \right) \;, \\
\Sigma^{h_1} &= (\mu^{s^2} + \Sigma^{s^2}) \cdot \left(\mathrm{trace}\{2(\mathbf{Q}\Sigma^p)^2\} + 4(\underline{\mu}^p)^T \mathbf{Q}\Sigma^p \mathbf{Q}\underline{\mu}^p \right) \;, \\
&\quad + \left(\mathrm{trace}\{\mathbf{Q}\Sigma^p + (\underline{\mu}^p)^T \mathbf{Q}\underline{\mu}^p\} \right) \cdot \Sigma^{s^2} \\
\Sigma^{h_1 m} &= \mathbf{0} \;, \\
\Sigma^{h_1 p} &= \mu^{s^2} \cdot \left(2\Sigma^p \mathbf{Q}\underline{\mu}^p + \mathrm{trace}\{\underline{q}(\phi)\}\Sigma^p \underline{q}(\phi) \right) \cdot \underline{\mu}^p + \underline{\mu}^p \cdot (\underline{\mu}^p)^T \mathbf{Q}\underline{\mu}^p \;,
\end{aligned}
$$

with $\mathbf{Q} := \underline{q}(\hat{\phi}^T) \underline{q}(\hat{\phi})$.

The moments with $\boldsymbol{h}_2$ turn out to be

$$
\begin{aligned}
\underline{\mu}^{h_2} &= (\underline{\mu}^m - \hat{\underline{y}})^T (\underline{\mu}^m - \hat{\underline{y}}) + \mathrm{trace}\{\Sigma^m\} \;, \\
\Sigma^{h_2} &= 2\,\mathrm{trace}\{(\Sigma^m)^2\} + 4(\underline{\mu}^m - \hat{\underline{y}})^T \Sigma^m (\underline{\mu}^m - \hat{\underline{y}}) \;, \\
&\quad + \mathrm{trace}\{\Sigma^m\} + (\underline{\mu}^m - \hat{\underline{y}})^T (\underline{\mu}^m - \hat{\underline{y}}) - (\underline{\mu}^h)^2 \;, \\
\Sigma^{h_1 m} &= 2 \cdot \Sigma^m (\underline{\mu}^m - \hat{\underline{y}}) + \mathrm{trace}\{\Sigma^m\} \cdot \underline{\mu}^m - \underline{\mu}^m \,\mathrm{trace}\{\Sigma^m\} \;, \\
\Sigma^{h_1 p} &= \mathbf{0} \;.
\end{aligned}
$$

Finally, the moments for the noise term $\boldsymbol{h}_3$ are

$$
\begin{aligned}
\underline{\mu}^{h_3} &= \mathrm{trace}\{\Sigma^v\} \;, \\
\Sigma^{h_3} &= \mathrm{trace}\{\underline{t}(\underline{t})^T \Sigma^v\} + 2\,\mathrm{trace}\{(\Sigma^v)^2\} + \mathrm{trace}\{\Sigma^v\}^2 \;, \\
\Sigma^{h_1 m} &= \mathbf{0} \;, \\
\Sigma^{h_1 p} &= \mathbf{0} \;,
\end{aligned}
$$

where $\underline{t} := 2 \cdot \mu^s \cdot \underline{q}(\hat{\phi}) \cdot \underline{\mu}^p \cdot \underline{e}(\hat{\phi})^T$.

The Gaussian filter using analytic moment calculation according to the above formulas is denoted as *AMC-SC-RHM* in the remainder of this work.

Unscented Kalman filter (UKF-SC-RHM)

The required quantities for performing the measurement update with (4.17), i.e., the mean $\underline{\mu}^h$, covariance matrix Σ^{hh}, and the cross-covariance matrix

Σ^{hx} can be computed approximately with the unscented transform [JU04]. We denote the resulting filter as *UKF-SC-RHM*. Although the unscented transform only gives an approximation of the moments, it has the advantage that one can resort to a standard nonlinear estimator for which optimized implementations are available. Simulations in Section 4.3.5 show that the performance of the *UKF-SC-RHM* is close to the filter based on analytic expressions *AMC-SC-RHM*. Hence, the *UKF-SC-RHM* is a serious alternative to the *AMC-SC-RHM*.

4.3.4 Particle Filter

As already discussed earlier in Section 2.3, particle filters for extended object tracking are rather challenging as, among others, there are no general analytic expressions for the likelihood function (2.7). In the following, we show that the concept of an *RHM* can be used for deriving an accurate approximation of the likelihood function that can be calculated efficiently. Essentially, this approximation exploits that the likelihood function can be evaluated explicitly when the angle from the center to the measurement source is given. Hence, only a one-dimensional discretization has to be performed (for two-dimensional shapes). This approximation is possible for all star-convex shapes, regardless of the parameterization.

With $\underline{z}_{k,l} = s_{k,l} \cdot r(\underline{p}_k, \phi_{k,l}) + \underline{m}_k$, we can rewrite (2.7) as

$$
\begin{aligned}
& \int f(\underline{\hat{y}}_{k,l}|\underline{z}_{k,l}) \cdot f(\underline{z}_{k,l}|\underline{x}_k) \, \mathrm{d}\underline{z}_{k,l} \\
= & \int \int f(\underline{\hat{y}}_{k,l}|\phi_{k,l}, s_{k,l}, \underline{x}_k) \cdot f(\phi_{k,l}, s_{k,l}, |\underline{x}_k) \, \mathrm{d}\phi_{k,l} \, \mathrm{d}s_{k,l} \\
= & \int \int f(\underline{\hat{y}}_{k,l}|\phi_{k,l}, s_{k,l}, \underline{x}_k) \cdot f(\phi_{k,l}|\underline{x}_k) \cdot f(s_{k,l}, |\underline{x}_k) \, \mathrm{d}\phi_{k,l} \, \mathrm{d}s_{k,l} \\
= & \int \underbrace{\left(\int f(\underline{\hat{y}}_{k,l}|\phi_{k,l}, s_{k,l}, \underline{x}_k) \cdot f(s_{k,l}|\underline{x}_k) \, \mathrm{d}s_{k,l} \right)}_{:= f^s(\underline{\hat{y}}_{k,l}|\phi_{k,l}, \underline{x}_k)} \cdot f(\phi_{k,l}|\underline{x}_k) \mathrm{d}\phi_{k,l} \quad (4.18)
\end{aligned}
$$

Note that we exploited that the scaling factor and the angle are independent. The term $f^s(\underline{\hat{y}}_{k,l}|\phi_{k,l}, \underline{x}_k)$ can be evaluated analytically for several relevant densities $f(s_{k,l}|\underline{x}_k)$, e.g., Gaussian and uniform densities (it results from (4.14) for fixed angle $\phi_{k,l}$).

For example, when $s_{k,l}$ is Gaussian distributed with mean μ^s and variance Σ^s, we obtain

$$f^s(\hat{\underline{y}}_{k,l}|\phi_{k,l},\underline{x}_k) = \mathcal{N}(\hat{\underline{y}}_{k,l} - \underline{\mu}^y, \Sigma^y) \ , \qquad (4.19)$$

where

$$\underline{\mu}^y \ = \ \mu^s \cdot r(\underline{p}_k, \phi_{k,l}) \cdot \underline{e}(\phi_{k,l}) + \underline{m}_k \ , \qquad (4.20)$$

$$\Sigma^y \ = \ \Sigma^v_{k,l} + r(\underline{p}_k, \phi_{k,l})^2 \cdot \underline{e}(\phi_{k,l})^T \Sigma^s \underline{e}(\phi_{k,l}) \ . \qquad (4.21)$$

Because the remaining outer integral in (4.18) cannot be solved in closed form, we suggest to approximate the probability density $f^L(\phi_{k,l}|\underline{x}_k)$ with a Dirac mixture distribution

$$f^L(\phi_{k,l}|\underline{x}_k) \approx \sum_{i=1}^{N_\phi} w_i \delta(\phi_{k,l} - \phi_{k,l}^{(i)})$$

With this approximation, we obtain the following approximation of the likelihood (2.7)

$$f(\hat{\underline{y}}_{k,l}|\underline{x}_k) \ \approx \ \sum_{i=1}^{N_\phi} w_i f^s(\hat{\underline{y}}_{k,l}|\phi_{k,l}^{(i)},\underline{x}_k) \ , \qquad (4.22)$$

Finally, it remains open how to choose the weights w_i. When the measurement sources are uniformly distributed over the object surface, the angle $\phi_{k,l}$ is in general *not* uniformly distributed as well. Instead, a more suitable approximation for the weights w_i is given by

$$w_i := \frac{||r(\underline{p}_k, \phi_{k,l}^{(i)}) - r(\underline{p}_k, \phi_{k,l}^{(i-1)})|| + ||r(\underline{p}_k, \phi_{k,l}^{(i)}) - r(\underline{p}_k, \phi_{k,l}^{(i+1)})||}{2 \cdot C}$$

for $1 \leq i \leq N_\phi$, where we define $\phi_{k,l}^{(N_\phi+1)} := \phi_{k,l}^{(1)}$ and $\phi_{k,l}^{(-1)} := \phi_{k,l}^{(N_\phi)}$. The term C denotes an approximation of the curve length of star-convex boundary is given by

$$C \approx \frac{1}{N_\phi} \sum_{i=1}^{N_\phi-1} ||r(\underline{p}_k, \phi_{k,l}^{(i)}) - r(\underline{p}_k, \phi_{k,l}^{(i+1)})|| \ .$$

Essentially, this approximation results from approximating the shape with a "fan", i.e., the shape boundary is approximated with line segments, and

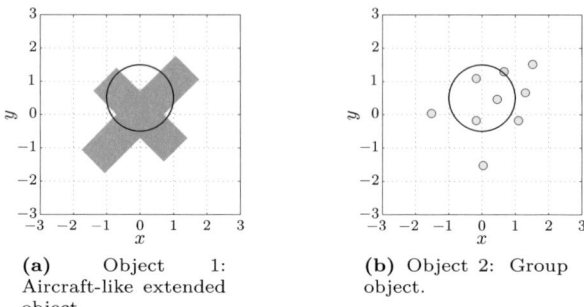

(a) Object 1: Aircraft-like extended object.

(b) Object 2: Group object.

Figure 4.9: Extended objects in the evaluation and the prior shape estimate, i.e., a circle.

the weight w_i is chosen as the ratio between the two adjacent segments of $\phi_{k,l}^{(i)}$ to the overall area. The above proposed approximation can be performed in linear runtime in N_ϕ. Having approximated the likelihood function, standard particle filters can be used for tracking the extended object.

We introduce the acronym *PF-SC-RHM* for the particle filter that is equal to the *Naïve-PF* described in Section 2.3 but employs the above described approximation of the likelihood.

4.3.5 Evaluation

The performance of the proposed methods for star-convex shapes is investigated within two scenarios. The first scenario considers a stationary extended object with fixed shape. The purpose of this scenario is twofold. First, it shows the practicability of star-convex *RHMs* and second, the performance of the two Gaussian filter implementations *UKF-SC-RHM* and *AMC-SC-RHM* are compared with both the naïve particle filter and *PF-SC-RHM*. The second scenario shows the applicability of the proposed methods for extracting detailed shape information from a moving extended object.

Stationary Extended Object

Simulations for the first scenario are carried out with two extended objects, i.e., an aircraft-like object and a group object as depicted in Fig. 4.9 (from both objects, measurement sources are drawn uniformly). From each object, 200 measurements are sequentially received, where simulations are performed under different measurement noise levels, i.e.,

- low measurement noise level $\Sigma_{k,1}^v = \mathrm{diag}(0.1^2, 0.1^2)$,

- medium measurement noise level $\Sigma_{k,1}^v = \mathrm{diag}(0.27^2, 0.27^2)$, and

- high measurement noise level $\Sigma_{k,1}^v = \mathrm{diag}(0.4^2, 0.4^2)$.

Example measurements for all noise levels are depicted for both objects in Fig. 4.10 and Fig. 4.12.

The shape of the object is represented with the polar representation using 11 Fourier coefficients as described in Section 4.3.1. The squared scaling factor for the *RHM* is Gaussian distributed with mean 0.5 and variance $\frac{1}{12}$. The parameters of the shape are initialized with a Gaussian with mean $[0.5, 0.5, 2, 0, \ldots, 0]^T$ and covariance $\mathrm{diag}(0.2, 0.2, 0.04, 0.04, \ldots, 0.04)$, i.e., an uncertain circle with radius 1 and center $[0, 0.5]^T$.

AMC-SC-RHM   First, the shape of the stationary extended object is recursively estimated with the closed-form implementation *AMC-SC-RHM* of star-convex *RHMs*. The resulting point estimates after 50 and 200 measurements are depicted in Fig. 4.11 and Fig. 4.13, where they are averaged over 20 runs. In order to visualize the estimation accuracy, we have computed an (empirical) "variance" for the shape estimate. This shape variance results from the 2σ-bound of the radius function, i.e., for each ϕ the variance for $r(\underline{p}_k^{\mathrm{sc}}, \phi)$ is computed based on the 50 runs.

The results demonstrate that the shape is estimated precisely for both shapes. Even the shape of the group object is captured quite well although there is a significant modeling mismatch between the group object and the assumptions of the *RHM*. Furthermore, with an increasing measurement noise, the details of the shape become blurred. This is a reasonable effect because with lower sensor resolution, details are more difficult to extract and a higher number of measurements is necessary.

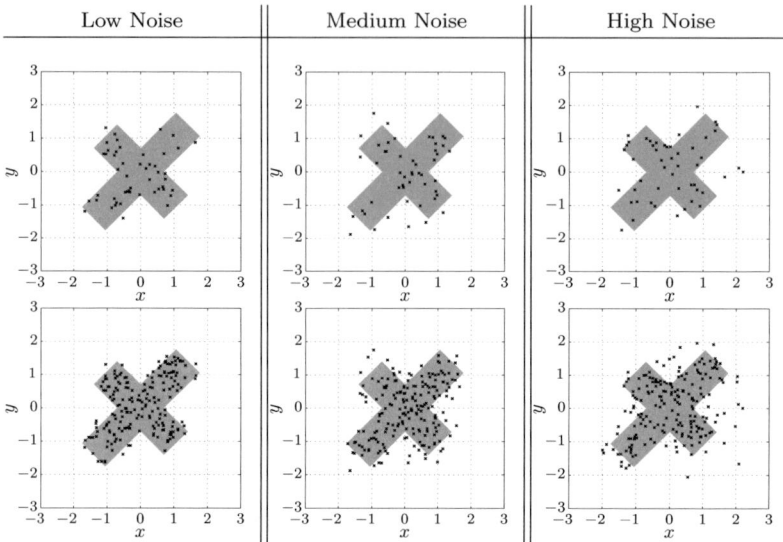

Figure 4.10: Object 1: 200 example measurements (second row) and the first 50 measurements (first row).

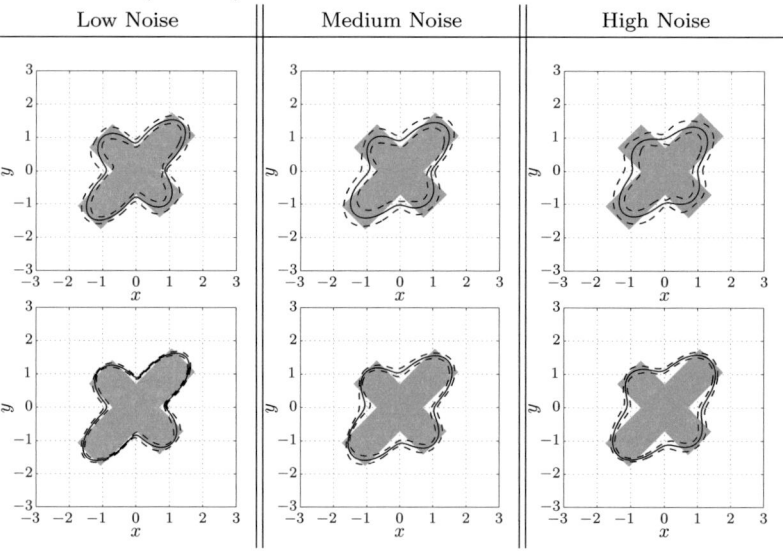

Figure 4.11: Object 1: Shape estimates averaged over 20 runs after 50 and 200 measurements.

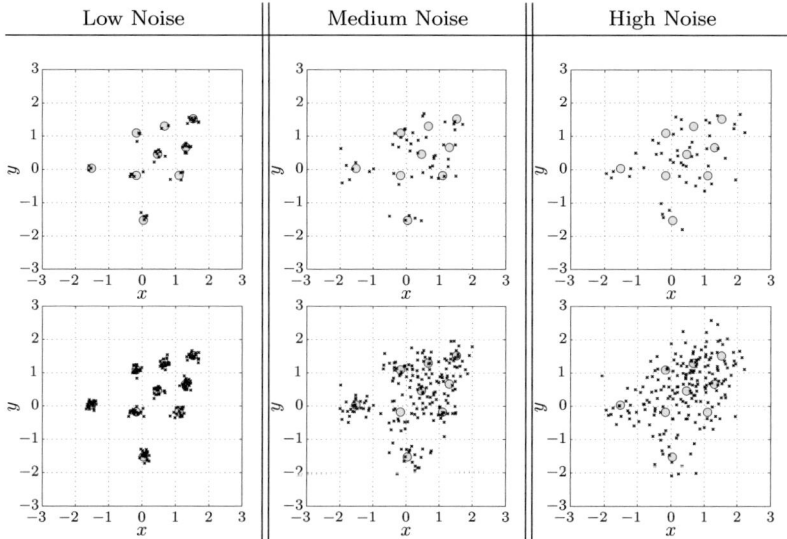

Figure 4.12: Object 1: 200 example measurements (second row) and the first 50 measurements (first row).

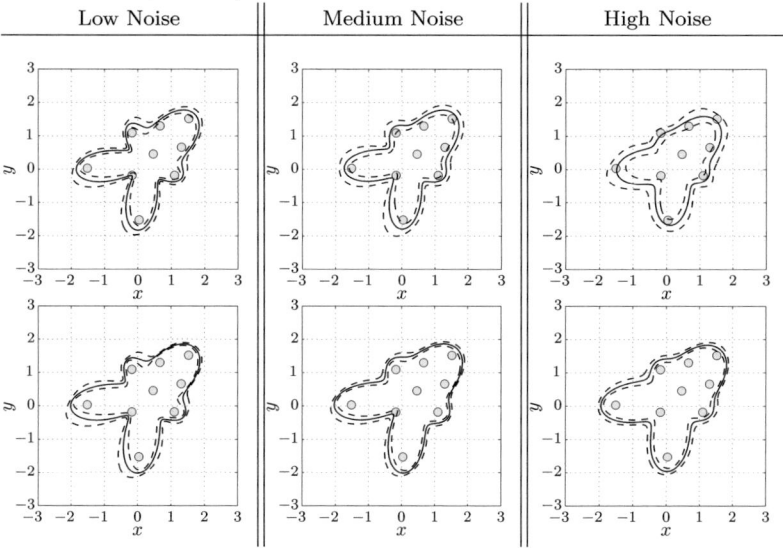

Figure 4.13: Object 1: Shape estimates averaged over 20 runs after 50 and 200 measurements.

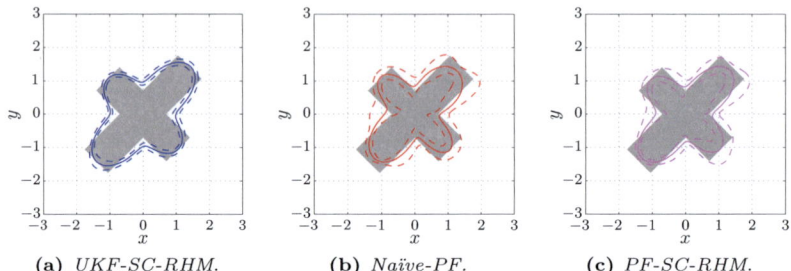

(a) *UKF-SC-RHM.* (b) *Naïve-PF.* (c) *PF-SC-RHM.*

Figure 4.14: Comparisons with the help object 1 and middle noise level: Mean shape estimates and shape variance based on 50 runs.

AMC-SC-RHM vs. UKF-SC-RHM In general, we observed that the *UKF* implementation *UKF-SC-RHM* and the closed-form solution *AMC-SC-RHM* result in comparable shape estimates. In order to support this statement, the estimation results of the *UKF-SC-RHM* for the extended object Fig. 4.9a and middle measurement noise level are shown in Fig. 4.14a.

AMC/UKF-SC-RHM vs. Particle filters Within this thesis, we introduced two particle filter implementations, i.e., the naïve particle filter *Naïve-PF* (see Section 2.3) and a particle filter *PF-SC-RHM*, which equals to the naïve particle filter except that the concept of an *RHM* is used for deriving an approximation of the likelihood function.

For both particle filters, 1000 samples were employed for representing the probability density of the state. For the *Naïve-PF*, the likelihood function was approximated based on 500 samples (note that the naïve particle filter performs a two-dimensional sampling of the object surface). However, the *PF-SC-RHM* employed only 250 samples (for the one-dimensional angle in the likelihood function). The number of samples was chosen as high as possible while resulting in a reasonable run-time (for each measurement update 500.000 samples have to be iterated for the naïve particle filter). Of course, the run-time of the analytic approach *AMC-SC-RHM* is unbeatable. However, the *UKF* implementation is only slightly slower.

Fig. 4.14b and Fig. 4.14c show the estimation results for the both particle filter implementations. The *Naïve-PF* and *PF-SC-RHM* have a much larger error than both Gaussian filter implementations *UKF-SC-RHM* and

AMC-SC-RHM. Although the *PF-SC-RHM* employs less samples than the naïve particle filter the estimation quality is comparable. This results from the fact that *PF-SC-RHM* approximates the likelihood more precisely with the same number of particles than the *Naïve-PF.*

All told, the evaluation shows that the Gaussian filter implementations are hard to beat with basic particle filter techniques concerning both run-time and estimation accuracy. Of course, more advanced particle filtering techniques probably give more competitive results.

Tracking a Moving Extended Object

In the second scenario, a star-convex shape approximation of a moving extended object is to be tracked (see Fig. 4.15a). The extended object follows the trajectory depicted in Fig. 4.15b, where the measurement sources are uniformly distributed on the object's surface. This is exactly the same object and trajectory as for ellipses in Section 4.2.4. The magnitude of the measurement noise is nearly the same as in Section 4.2.4, i.e., the covariance matrix of the measurement noise is $\Sigma_{k,1}^v = \mathrm{diag}(0.2, 0.2)$ with probability 0.6 and $\Sigma_{k,1}^v = \mathrm{diag}(0.6, 0.6)$ with probability 0.4. However, in contrast to Section 4.2.4, the number of measurements received per time instant is increased, i.e., $n_k = n_k^* + 1$, where n_k^* is Poisson distributed with mean 7.

The shape of the object is tracked with *UKF-SC-RHM* implementation for star-convex extended objects, where a Gaussian scaling factor with mean 0.7 and variance 0.05 is employed for the *RHM*. The state vector is $\underline{x}_k^{\mathrm{sc}} = \left[\underline{m}_k^T, (\underline{m}_k^v)^T, (\underline{p}_k^{\mathrm{sc}})^T \right]^T$, where $\underline{m}_k$ is the center, $\underline{m}_k^v$ is the velocity vector, and $\underline{p}_k^{\mathrm{sc}}$ are the shape parameters consisting of 11 Fourier coefficients. We use the motion model (2.5) described in Section 2.1.3, where the center evolves according to constant velocity model and shape parameters are modeled as a random walk. The noise parameters for (2.5) are $\mathbf{C}_k^p = 0.001 \cdot \mathbf{I}_{11}$ and $\mathbf{C}_k^{cv} = 0.005 \cdot \mathbf{I}_2$.

The estimated star-convex shape approximation is shown in Fig. 4.15 for two details of the overall trajectory (averaged over 20 time steps). The results indicate that the shape is tracked precisely, i.e., the aircraft-like shape can be recognized. However, similar to the elliptic shape in Section 4.2.4, the shape marginally blurs as soon as the extended object changes its orientation.

Of course, the object in this scenario could also be tracked with an elliptic model (see Section 4.2.4). However, an ellipse does not encode the aircraft-like shape of the object. Vice versa, the object in the scenario for elliptic shapes in Section 4.2.4 cannot be tracked with a star-convex *RHM* using 11 Fourier coefficients as the measurement quality in Section 4.2.4 is too low, i.e., the measurement noise is too high, too few measurements are available, and the object alters its moving direction too quickly. Nevertheless, the object in Section 4.2.4 can be tracked with a star-convex *RHM* when the number of Fourier coefficients is reduced, which, however, comes with a loss of the detailed shape information.

4.4 Conclusions

In this chapter, we considered the tracking of an extended object that is modeled as a region shape, i.e., the point measurements may originate from the *interior* of the object contour. We faced the high-dimensionality and nonlinearity of the estimation problem with a novel concept called *Random Hypersurface Model* (*RHM*) that allows for reducing the modeling of a region shape to the modeling of a curve. This reduction is performed by means of scaling the shape boundary according to a scaling factor that is characterized with a one-dimensional probability density function. *RHMs* offer a flexible concept for modeling free-form star-convex shapes but also basic shapes, e.g., ellipses. We introduced particular *RHM* instantiations for elliptic shapes that are represented with an implicit equation and star-convex shapes represented with an explicit polar function. For both shapes, quadratic measurement equations were derived for which statistical linearization yields an efficient closed-form measurement update. The estimation accuracy and performance was highlighted with respect to basic particle filters. For this purpose, the concept of an *RHM* was also used for deriving accurate approximations of the extended object likelihood function.

At this point, it is essential to note that the contributions of this chapter are twofold: First, we presented the novel fundamental model called *RHM*

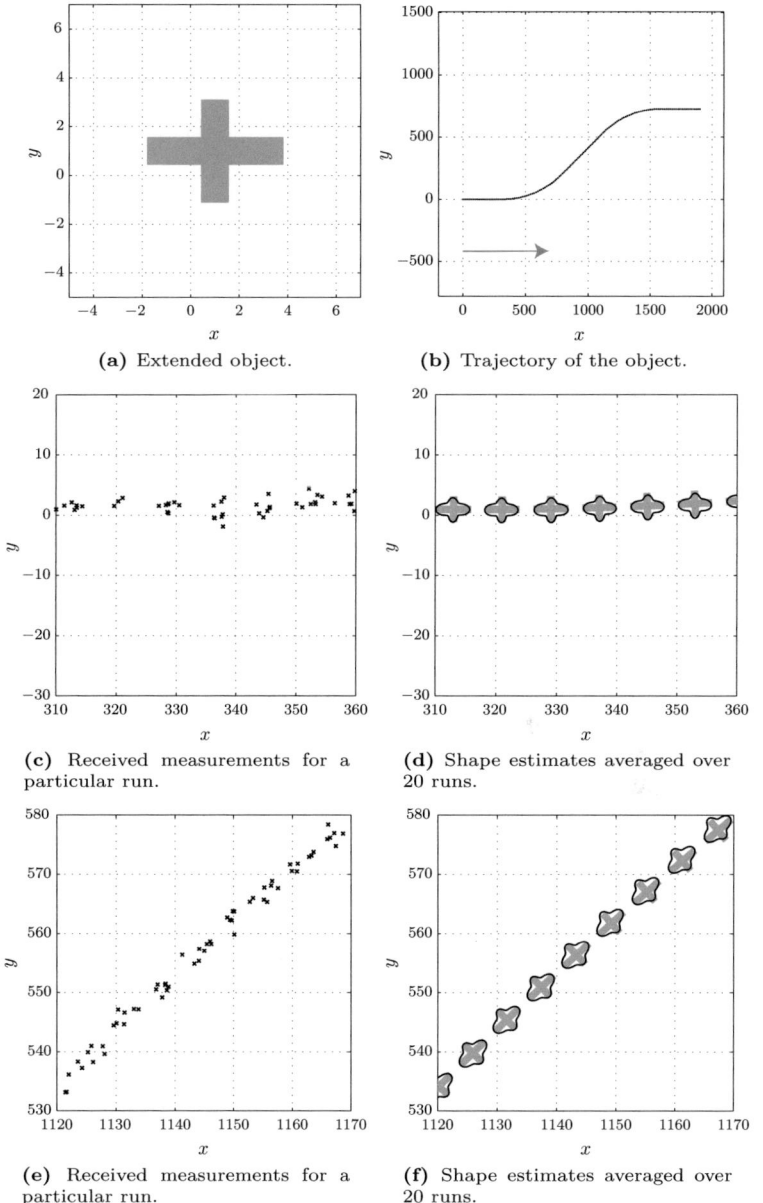

(a) Extended object.

(b) Trajectory of the object.

(c) Received measurements for a particular run.

(d) Shape estimates averaged over 20 runs.

(e) Received measurements for a particular run.

(f) Shape estimates averaged over 20 runs.

Figure 4.15: Tracking an extended object with an *RHM* for star-convex objects and a constant velocity model.

that allows for deriving explicit measurement equations for extended objects. Second, particular state estimators for these measurement equations have been derived.

This chapter demonstrated that it is possible to efficiently extract detailed shape information from sequentially arriving noisy point measurements. It is now possible to track an extended object whose shape is totally unknown a priori and estimated from scratch. Of course, this is only possible under certain good conditions, e.g., the measurement noise may not be too high. If the conditions are too poor, a coarse shape model such as an ellipse is recommended in order to prevent track loss.

The closed-form formulas for the measurement update render the presented methods computationally appealing for many applications. An example application will be given later in Chapter 6, where the real-time shape tracking of ground moving objects with *RHMs* is considered.

Outlook Concerning the proposed Gaussian filter instantiations of *RHMs*, there is room of improvements in the shape parameterization. The Fourier series expansion of the polar function comes with the disadvantage that there is an implicit non-negativity constraint, i.e., the polar function must be always positive. Furthermore, the center is only uniquely defined in connection with a particular distribution of the scaling factor. Both issues can be addressed by either enforcing state constraints or choosing alternative representations for star-convex shapes. As already discussed in the context of conic fitting, a Gaussian density for the center/shape representation of an ellipse does not have an intuitive physical meaning, e.g., it is unclear what the average of two ellipses is.

Of course, advanced, problem-specific non-Gaussian filters (e.g., based on an *RHM*) are expected to give even more precise shape estimates than the proposed analytic Gaussian solution. However, non-Gaussian filters usually have their price; they are in general computationally far more expensive. The question is how much more estimation quality can be achieved with how much effort. The comparison with the basic particle filters indicates that our analytic approach is hard to beat in both quality and run-time complexity.

In this chapter, we have focused on the basic version of *RHMs* for star-convex shapes. However, the concept of an *RHM* is much more general. For example, the following two generalizations are promising:

- *More complex shapes*
  Although a star-convex shape is a very precise description of an object, there may be scenarios in which non-star-convex shapes are required. There are actually two main approaches that allow for extending *RHMs* to non-star-convex shapes. First, a complex shape can be composed of several basic shapes. By this means, a data association problem arises as it is not known from which shape a measurement stems (see [19] for preliminary results on this approach). Second, the idea of scaling the shape boundary can be generalized to arbitrary shapes. While the scaling corresponds to a straight-line homotopy from the shape to the center, it is possible to define a homotopy for more general shapes in a similar manner.

- *Three-dimensional space*
  Elliptic region shapes can be treated in higher dimensions in analogy to the two-dimensional case. However, star-convex region shapes in three-dimensional space are somewhat more difficult as a two-dimensional Fourier expansion has to be performed. Apart from region shapes, i.e., three dimensional shapes, *RHMs* can be used for modeling *surfaces*. For example, in [22], *RHMs* were used for tracking and estimating the parameters of a cylinder. Tracking the parameters of a surface in three-dimensional space is a very important problem that recently gained a lot of attention in the context of processing point clouds obtained from RGB and depth sensors.

CHAPTER 5

Set-Theoretic Extent Model

Contents

Existing extended object tracking approaches impose statistical assumptions on the measurement sources, e.g., a spatial distribution may assume that the measurement sources are uniformly distributed on the object. However, statistical assumptions are often unreasonable as the locations of the measurement sources are influenced by effects that are hard or even impossible to identify. For example, the measurement sources may be significantly affected by unknown characteristics of the object's surface, the sensor-to-object geometry, or the arrangement of the individual group members in a group. Definitely, unreasonable assumptions on the measurement sources may sooner or later result in a low estimation quality. The plain Bayesian solution would be to estimate the parameters of the spatial distribution. However, the Bayesian approach comes with two disadvantages. First, it would be computationally challenging as further unknown parameters are to be estimated. Second, one would also need models for the temporal evolution of these parameters, which are hard to justify, e.g., it would be necessary to explicitly model occlusions. In this chapter, a novel approach is pursued:

The location of a measurement source is modeled as an
unknown-but-bounded error.

Hence, it is only exploited that a measurement source lies on the extended object, but no statistical assumptions are imposed. By this means, no wrong assumptions on the measurement sources are made.

At this point, it is essential to note that the measurement noise itself is modeled as a stochastic noise term according to the sensor model described in Section 2.1.2; only the uncertainty about the measurement source, i.e., the extent model, is modeled as a set.

In order to perform inference based on the set-theoretic extent model, we adopt the *Statistical and Set-theoretic Information (SSI) Filter* [HH99d, HH99a, HHS99, HH99b, HH99c] framework, which employs random set intersection for fusion. Specifically, we present novel outer-bounding techniques for tracking circular discs. A direct consequence of the imposed set-theoretic extent model is that the shape parameters, i.e., the radius, cannot be estimated exclusively with point measurements. However, it can be estimated from further information sources, e.g., we propose to infer the radius of the circular disc based on the number of received measurements.

The *SSI filter* for the set-theoretic extent model is robust against different distributions of the measurement sources and is thus able to give more precise estimation results than plain stochastic approaches. It turns out that a set-theoretic extent model may be superior to a spatial distribution model in case the locations of the measurement sources are dominated by systematic errors and the measurement noise is rather small compared to the extent.

Remark 5.1. This chapter is based on [2] and [8,10]. In [8], the set-theoretic approach has first been published. An extended and revised journal version was published in [2]. The approach was applied to rectangular shapes in [10]. This chapter extends the previous work [2] and [8,10] mainly by a motivating example and further evaluations.

Structure of this Chapter In the following section, we illustrate the need of a set-theoretic extent model by means of a simple example. Subsequently, the basic concept of the set-theoretic extent model is introduced in Section 5.2. Section 5.3 shows how the so-called *Statistical and Set-theoretic Information (SSI) Filter* can be used for performing inference based on the set-theoretic extent model. In Section 5.4, we derive a particular *SSI filter* for circular discs when the radius is known. For the case of unknown radius, novel outer-bounding techniques based on hyperboloids are developed in Section 5.5. In order to estimate the radius, it is assumed that the number of received measurements depends on the size of the object (see Section 5.5). The benefits of the set-theoretic approach are illustrated by means of simulations in Section 5.6.

5.1 Motivating Example

In order to illustrate the effect of statistical assumptions on the measurement sources, we consider the non-moving, cross-shaped extended object depicted in Fig. 5.1a. The measurements are received sequentially from a finite set of (equally probable) measurement sources on the object (see Fig. 5.1a). The goal is to estimate the location of the extended object. For this purpose, the extended object is modeled as a circular disc, where the radius of the smallest enclosing circle is known. A reasonable stochastic estimator for the center can be derived easily as follows.

Remark 5.2 (Stochastic estimator for discs with known radius). For given radius, the assumption that the measurement sources are uniformly distributed on the circular disc results in the measurement equation

$$\underline{\hat{y}}_{k,l} = \underline{m} + \underline{e}_{k,l} + \underline{v}_{k,l} \ ,$$

where $\underline{e}_{k,l}$ is a random variable that is uniformly distributed on a disc with radius $\tilde{r}_k$ at the origin denoted as $\mathbf{K}\left([0,0]^T, \tilde{r}_k\right)$. When approximating the uniform distribution of $\underline{e}_{k,l}$ with a Gaussian, the standard Kalman filter formulas can be employed for performing a measurement update as $\underline{e}_{k,l}$ is just an additional noise term in a linear model. Reasonably, a uniform distribution can be approximated with a Gaussian by moment matching. However, as in this case a large amount of the probability mass for a measurement source lies outside of the circular disc, we propose to choose the variance of the Gaussian so that the 0.95% confidence region is a circle with radius $\tilde{r}_k$.

The discussed example is a typical scenario, where the assumed spatial distribution, i.e., a uniform distribution, significantly differs from the true one, i.e., a finite set of point sources. Fig. 5.1 shows an example run of the stochastic estimator, which demonstrates the consequence of the modeling mismatch. Fig. 5.1b depicts the estimated circle, i.e., its center, and the 0.95% confidence ellipse after the first measurement has been processed. It can be seen that the true center is still contained in the confidence region.

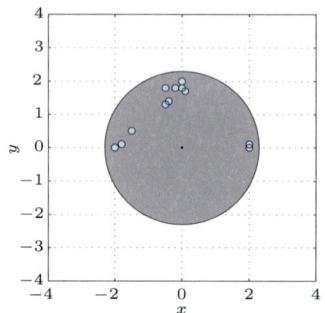

(a) Extended object with measurement sources.

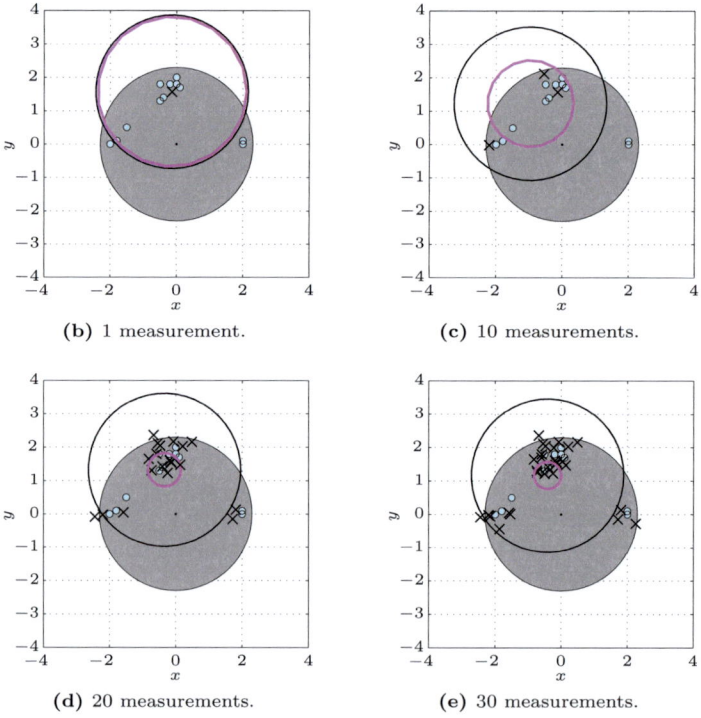

(b) 1 measurement.

(c) 10 measurements.

(d) 20 measurements.

(e) 30 measurements.

Figure 5.1: Example illustrating the problem of systematic errors in the measurement sources when using a stochastic estimator (see Example 5.1). The confidence ellipse of the stochastic estimator is plotted in magenta.

However, as soon as the number of processed measurements increases (see Fig. 5.1d and Fig. 5.1e), the estimated center drifts away and the confidence region does not contain the true center anymore. Actually, this is a direct consequence of the modeling mismatch and results in a systematic error. Essentially, the estimated center coincides with the mean of the received measurements. However, the center of the object does not coincide with the mean of the measurement sources. The misbehavior of the stochastic estimator may cause serious problems in a tracking system as the estimator pretends to be more confident than it is actually the case. In the following, we call such an estimator inconsistent (see for example [LBDS04]).

Definition 5.1 (Inconsistency). An estimator is called *inconsistent* if it considers the true state as an outlier, i.e., the confidence region does not contain the true state.

5.2 Basic Idea

The basic idea for coping with the problems described in the motivating example is to model the uncertainty about the measurement source as an unknown-but-bounded error, i.e., the only used information about the measurement source is that it lies on the extended object. For this purpose, we first restrict ourselves to the case that the shape parameters $\tilde{\underline{p}}_k$ are known. Hence, the state vector to be estimated $\underline{x}_k = \underline{m}_k$ consists only of the center $\underline{m}_k$ (for the sake of simplicity, we omit further variables for the kinematics).

The set-theoretic extent model is specified by the measurement equation

$$\hat{\underline{y}}_{k,l} = \underbrace{\underline{m}_k + \underline{e}_{k,l}}_{=\underline{z}_{k,l}} + \underline{v}_{k,l} \ , \qquad (5.1)$$

where

- $\underline{v}_{k,l}$ is zero-mean white Gaussian noise with covariance matrix $\Sigma^v_{k,l}$ that models the measurement error stemming from the sensor (see Section 2.1.2), and

- $\underline{e}_{k,l}$ is an unknown-but-bounded error term, i.e.,

$$\underline{e}_{k,l} \in \mathcal{S}(\tilde{\underline{p}}_k) \ ,$$

  which models the uncertainty of the measurement source on the object (see the extent model in Section 2.1.2).

Remark 5.3. Equation (5.1) indicates that $\tilde{\underline{p}}_k$ cannot be estimated based on the measurements $\hat{\underline{y}}_{k,l}$, as $\tilde{\underline{p}}_k$ specifies the bound of a systematic error. It is not possible to estimate it based on repeated observations $\hat{\underline{y}}_{k,l}$. In order to deal with this problem, we will later slightly extend the basic model proposed in Chapter 2: The number of measurements n_k received at a time instant depends on the size of the object. By this means, it is at least possible to estimate the scaling of the object. Further details, including a justification of this extended model, are provided later in Section 5.5.

5.3   Statistical and Set-theoretic Information (SSI) filter for Extended Objects

In this thesis, we will employ a *Statistical and Set-theoretic Information (SSI) Filter* [HH99a, HHS99, HH99b, HH99c, HH99d] for state estimation based on the set-theoretic extent model (5.1). An *SSI filter* provides a mathematically sound formalism for dealing with combined uncertainties, i.e., stochastic and set-theoretic uncertainties.

An *SSI filter* based on the measurement equation (5.1) represents the uncertainty about the state $\underline{x}_k$ having received the measurements up to time step $k-1$ plus the first $l-1$ measurements $\hat{\underline{y}}_{k,1}, \ldots, \hat{\underline{y}}_{k,l-1}$ from time step k with a random set $\Delta_{k,l-1}$, which is called *solution set*.

In order to incorporate the next measurement $\hat{\underline{y}}_{k,l}$ into the current solution set $\Delta_{k,l-1}$, the measurement equation (5.1) is written as

$$\underline{m}_k \in \underbrace{\hat{\underline{y}}_{k,l} - \mathcal{S}(\tilde{\underline{p}}_k) - \underline{v}_{k,l}}_{:=\Theta_{k,l}} \ , \tag{5.2}$$

which means that all feasible centers $\underline{m}_k$ that are consistent with the measurement $\hat{\underline{y}}_{k,l}$ are an element of a random set $\Theta_{k,l}$. Note that the set $\Theta_{k,l}$ is random, because $\underline{v}_{k,l}$ is a random vector.

The updated solution set $\Delta_{k,l}$ is the intersection of the current solution set $\Delta_{k,l-1}$ with the measurement solution set $\Theta_{k,l}$ (conditioned on non-emptiness), i.e.,

$$\Delta_{k,l} = \Delta_{k,l-1} \cap \Theta_{k,l} \ .$$

Since the exact recursive computation of $\Delta_{k,l}$ is in general intractable, the set $\Delta_{k,l}$ can be chosen to be an outer-bound of the true solution set, i.e.,

$$\Delta_{k,l} \supseteq \Delta_{k,l-1} \cap \Theta_{k,l} \ .$$

Typical sets are random intervals [HHS99] or random ellipsoids [HH99d].

Remark 5.4. On the one hand, the *SSI filter* becomes a pure set-theoretic filter if there is no stochastic noise. On the other hand, if the bounded error is zero, a stochastic filter is obtained.

Remark 5.5. Note that there are several further approaches for dealing with combined set-theoretic and stochastic uncertainties. For example, in the context of random set theory, Mahler [Mah07] suggested to construct a so-called *generalized likelihood function*, which incorporates both stochastic and set-theoretic errors. The uncertainty of the state is still represented with a probability density function for the state. In 1991, sets of densities [MS91, MS03] for a *Set-valued Kalman filter* were proposed. These ideas have been recently further developed in [NKH09].

In order to use an *SSI filter* for extended objects, outer-bounding techniques tailored to the particular error bound $\mathcal{S}(\underline{\tilde{p}}_k)$ have to be developed. In the following, we derive these techniques for circular discs, which are one of the most relevant shapes. Although, the set-theoretic approach is applicable for general shapes (see also Section 5.7), it makes mainly sense for coarse shapes, because the shape parameter cannot be estimated from point measurements. Still, a coarse shape outer-bounds a detailed shape and therefore guarantees consistent estimation results.

5.4 Circular Discs with Known Radius

In this section, we focus on an extended object modeled as a circular disc with known radius $\tilde{r}_k$, i.e., $\underline{\tilde{p}}_k = \tilde{r}_k$, and shape $\mathcal{S}(r_k) := \mathbf{K}(\tilde{r}_k)$, where $\mathbf{K}(\tilde{r}_k) := \{\underline{z} \mid \underline{z} \in \mathbb{R}^2 \text{ and } ||\underline{z}||_2 \leq \tilde{r}_k\}$. In this case, the measurement

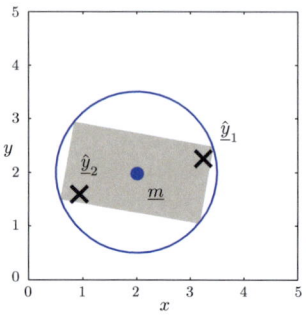

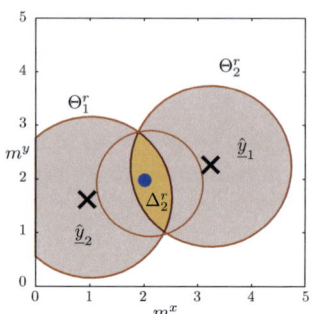

(a) A rectangular extended object plus its smallest enclosing circle and two position measurements $\hat{\underline{y}}_1$ and $\hat{\underline{y}}_2$.

(b) Intersection of the measurement solution sets for $\hat{\underline{y}}_1$ and $\hat{\underline{y}}_2$ bounded by a circle.

Figure 5.2: Set-theoretic estimation of the center of a circle with known radius.

solution set (5.2) is a circular disc with random center $\underline{z}_{k,l} := \hat{\underline{y}}_{k,l} - \underline{w}_k$ and given radius $\tilde{r}_k$, i.e.,[1]

$$\Theta^r_{k,l} := \mathbf{K}\left(\underline{z}_{k,l}, \tilde{r}_k\right) \ .$$

Example 5.1. Fig. 5.2a shows a rectangular shape with its smallest enclosing circle. The first two measurements $\hat{\underline{y}}_1$ and $\hat{\underline{y}}_2$ are plotted, where they are noise-free, i.e., $\underline{w}_1 = 0$ and $\underline{w}_2 = 0$ (the time index k is omitted in the example). The corresponding measurement solution sets for the center Θ^r_1 and Θ^r_2 are (deterministic) circular discs (see Fig. 5.2b). Because $\hat{\underline{y}}_1$ and $\hat{\underline{y}}_2$ both lie on the true circular disc, the center $\underline{m}_k$ is an element of $\Delta^r_2 = \Theta^r_1 \cap \Theta^r_2$.

The intersection of two measurement solution sets is not a circular disc again. However, the exact solution set can be outer-bounded with a circular disc (see Fig. 5.2b). In this manner, the solution set is given by

$$\Delta^r_{k,l} = \mathbf{K}\left(\underline{\xi}_{k,l}\right) \ ,$$

where $\underline{\xi}_{k,l} = \left[\xi^x_{k,l}, \xi^y_{k,l}, \xi^r_{k,l}\right]^T$ is a random vector consisting of the circle parameters. The intersection of the measurement solution set $\mathbf{K}\left(\underline{z}_{k,l}, \tilde{r}_k\right)$

---

[1]The superscript r in $\Theta^r_{k,l}$ expresses that the radius is known.

and the current solution set $\mathbf{K}\left(\underline{\xi}_{k,l-1}\right)$ has to be outer-bounded with its smallest enclosing circular disc according to

$$\mathbf{K}\left(\underline{\xi}_{k,l-1}\right) \cap \mathbf{K}\left(\underline{z}_{k,l}, \tilde{r}_k\right) \subset \mathbf{K}\left(\underline{\xi}_{k,l}\right) .$$

The following theorem shows how the updated parameters $\underline{\xi}_{k,l}$ can be computed from $\underline{\xi}_{k,l-1}$ and $\underline{z}_{k,l}$ (see also Fig. 5.2b).

Theorem 5.1. *Suppose we are given two circles* $\mathbf{K}\left(\underline{z}, \tilde{r}\right)$ *with* $\underline{z} = [z^x, z^y]^T$ *and* $\mathbf{K}\left(\xi\right)$ *with* $\xi = [\xi^x, \xi^y, \xi^r]^T$. *Let*

$$d := \sqrt{(z^x - \xi^x)^2 + (z^y - \xi^y)^2}$$

denote the distance between the vectors $[z^x, z^y]^T$ *and* $[\xi^x, \xi^y]^T$. *In case*

$$d \leq \xi^r + \tilde{r} \tag{5.3}$$

holds, the smallest enclosing circle $\mathbf{K}\left(\xi_{Bound}\right)$ *of the intersection* $\mathbf{K}\left(\underline{z}, \tilde{r}\right) \cap \mathbf{K}\left(\xi\right)$ *is given by*

$$\begin{bmatrix} \xi^x_{Bound} \\ \xi^y_{Bound} \end{bmatrix} := \underline{z} + \frac{c}{d} \cdot \left(\begin{bmatrix} \xi^x \\ \xi^y \end{bmatrix} - \underline{z} \right) , \tag{5.4}$$

$$\xi^r_{Bound} := \sqrt{\tilde{r}_k^2 - c^2} \tag{5.5}$$

with $c := 2((\tilde{r}^2 - \xi^r)^2 + d^2)$.

Proof. Can be proven with basic algebraic rules (see for example [Wei]). $\qquad\square$

According to the above discussion, the following *SSI filter* for the center of a circular disc can be constructed.

Statistical and Set-Theoretic Information (SSI) filter 1

- **State Vector**

$$\underline{\xi}_{k,l} \sim \mathcal{N}(\underline{\xi} - \hat{\underline{\xi}}_{k,l}, \Sigma^\xi_{k,l}) \text{ with } \underline{\xi}_{k,l} = \left[\xi^x_{k,l}, \xi^y_{k,l}, \xi^r_{k,l} \right]^T$$

- **Solution Set** for the center $\underline{m}_k$

$$\mathbf{K}\left(\underline{\xi}_{k,l}\right).$$

- **Measurement Solution Set** for measurement $\hat{\underline{y}}_{k,l}$
  $\mathbf{K}\left(\underline{z}_{k,l}, \tilde{r}_k\right)$ with $\underline{z}_{k,l} := \hat{\underline{y}}_{k,l} - \underline{w}_{k,l}$ and $\underline{z}_{k,l} \sim \mathcal{N}(y - \hat{\underline{y}}_{k,l}, \Sigma^v_{k,l})$.

- **Measurement Update**
  Calculate the parameters $\underline{\xi}_{k,l}$ of the outer-bound

$$\mathbf{K}\left(\underline{\xi}_{k,l}\right) \supseteq \mathbf{K}\left(\underline{\xi}_{k,l-1}\right) \cap \mathbf{K}\left(\underline{z}_{k,l}, \tilde{r}_k\right) \tag{5.6}$$

  by means of the mapping

$$\underline{\xi}_{k,l} = \mathcal{G}_1(\underline{z}_{k,l}, \underline{\xi}_{k,l-1}) \tag{5.7}$$

  under the condition that the intersection is not empty, i.e.,

$$\left\| \underline{z}_{k,l} - \left[\xi^x_{k,l-1}, \xi^y_{k,l-1}\right]^T \right\| \leq \tilde{r}_k + \xi^r_{k,l-1} \ , \tag{5.8}$$

  where the function $\mathcal{G}_1(\cdot)$ is defined according to Theorem 5.1.

- **Time Update**
  In general, the time update results from propagating the solution set through the motion model (2.3)

$$\Delta_{k+1,0} = a_k(\Delta_{k,n_k}, \underline{u}_k, \underline{w}_k) \cup \mathcal{E}_k \ ,$$

  where $\mathcal{E}_k$ is an additional bounded error for the prediction step. In case of a random walk model (2.4), the prediction step can be performed without any further approximations, i.e.,

$$\underline{\xi}_{k+1,0} = \begin{bmatrix} \mathbf{A}_k & 0 \\ 0 & 1 \end{bmatrix} \underline{\xi}_{k,n_k} + \begin{bmatrix} \hat{\underline{u}}_k + \underline{w}_k \\ 0 \end{bmatrix} + \begin{bmatrix} 0 \\ 0 \\ b_k \end{bmatrix} \ , \tag{5.9}$$

  where b_k is a scalar that accounts for $\mathcal{E}_k$ by increasing the radius of the solution set.

- **Point Estimate**
  A suitable point estimate is given by the expected center of the solution set $\left[\hat{\xi}^x_{k,l}, \hat{\xi}^y_{k,l}\right]^T$. An associated set-valued error is given by the radius of the solution set $\hat{\xi}^r_{k,l}$ and the stochastic error is given by the covariance matrix for $\left[\underline{\xi}^x_{k,l}, \underline{\xi}^y_{k,l}\right]^T$.

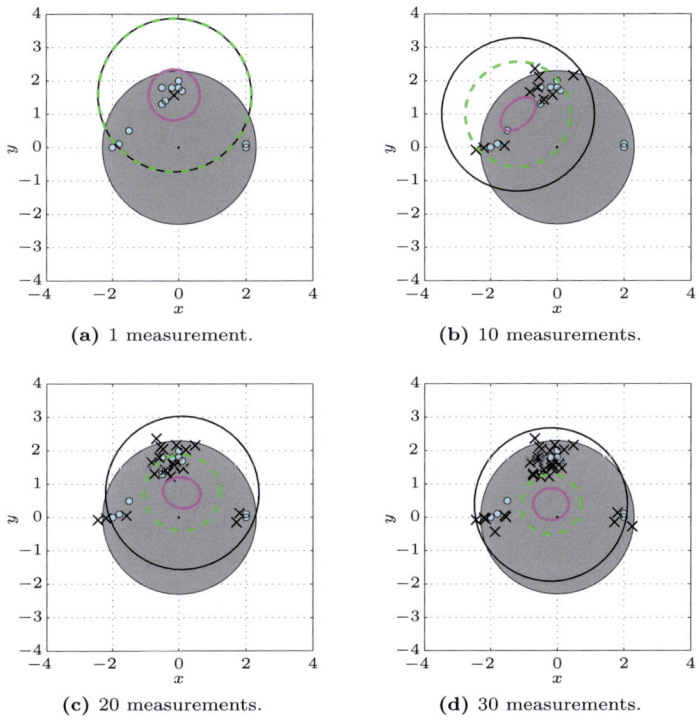

(a) 1 measurement.

(b) 10 measurements.

(c) 20 measurements.

(d) 30 measurements.

Figure 5.3: Estimation results of the *SSI filter* for the problem in Example 5.1. A state estimate is characterized by a confidence ellipse for the stochastic uncertainty (magenta) plus a bound for the set-theoretic uncertainty (dashed green).

The arguments of the nonlinear function $\mathcal{G}_1(\cdot)$ in (5.7) are random variables. In case $\underline{z}_{k,l}$ and $\underline{\xi}_{k,l-1}$ are Gaussian, $\underline{\xi}_{k,l}$ is not Gaussian anymore. However, a Gaussian approximation for the distribution of $\underline{\xi}_{k,l}$ can be maintained with the help of statistical linearization as introduced in Section 2.4, e.g., deterministic sampling approaches such as the *Unscented Kalman Filter* [JU04] are suitable. The state constraint (5.8) can easily be enforced when using deterministic sampling by dismissing or projecting infeasible samples.

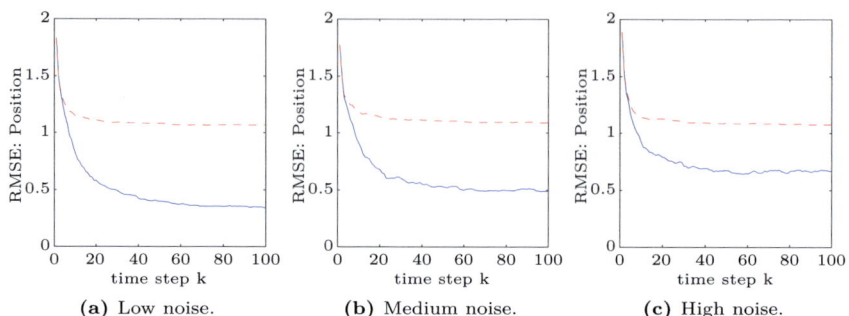

Figure 5.4: RMSE for the position: Stochastic estimator (red) vs. SSI (blue) filter, depending on the magnitude of the measurement noise (the noise levels are as in Section 5.6).

Motivating Example: Revisited The motivating example introduced in Section 5.1 can be revisited with SSI 1. In Fig. 5.3, SSI 1 has been used to estimate the center of the circular disc based on exactly the same measurements as the stochastic estimator in Fig. 5.1. As SSI 1 captures the uncertainty of the center with a random set, a stochastic confidence set and an expected bound for the set-valued error for the center can be determined. It can be seen in Fig. 5.3 that the set-valued error does not vanish totally with an increasing number of measurements, because the measurement sources do not cover the entire surface of the object. As a consequence, the true center is always contained in the overall confidence region specified by the random set (stochastic confidence region plus set-theoretic confidence region). All told, the SSI 1 is able to systematically treat the lack of knowledge about the measurement sources, which is also reflected in the *Root-Mean-Square Error* (*RMSE*) for the center shown in Fig. 5.4.

5.5 Circular Discs with Unknown Radius

In this section, the assumption of a known radius is dropped and an *SSI filter* for both the center $\boldsymbol{m}_k$ and radius $\boldsymbol{r}_k$ is constructed. For this purpose, the state vector is $\underline{\boldsymbol{x}}_k = \left[\boldsymbol{m}_k^T, \boldsymbol{r}_k \right]^T$ in the following.

Based on (5.1), we can infer that the measurement $\hat{\underline{y}}_{k,l}$ yields the measurement solution set

$$\left[\underline{m}_k^T, r_k\right]^T \in \left\{\left[\underline{m}^T, r\right]^T \in \Omega \ \Big| \ \|\underline{m} - \hat{\underline{y}}_{k,l} - \underline{v}_{k,l}\| \leq r^2\right\} =: \Theta_{k,l}, \quad (5.10)$$

of all vectors $\left[\underline{m}_k^T, r_k\right]^T$ that are consistent with $\hat{\underline{y}}_{k,l}$. In fact, $\Theta_{k,l}$ is a cone oriented along the r-axis with apex $\left[(\hat{\underline{y}}_{k,l} - \underline{v}_{k,l})^T, 0\right]^T$ and perpendicular cone angle.

Remark 5.6. As already discussed earlier, the radius r_k cannot be estimated based on point measurements in general. This results from the fact that it is in general not possible to estimate the magnitude of an unknown-but-bounded error (without imposing further assumptions). This becomes obvious when considering the case that each measurement originates from the center of the object, i.e., $\underline{z}_{k,l} = \underline{m}_k$ for all k and l. Later, we will propose to estimate the radius based on the number of received measurements and hence, model it as random variable. Note that the random solution set for the center then depends on all possible radii.

Example 5.2. Fig. 5.5a shows a rectangular shaped extended object from which two (noise-free) measurements $\hat{\underline{y}}_1$ and $\hat{\underline{y}}_2$ are received (the time index k is omitted). The corresponding measurement solution sets Θ_1 and Θ_2 are indicated in Fig. 5.5b. As $\hat{\underline{y}}_1$ and $\hat{\underline{y}}_2$ both lie on the circular disc, the parameter vector $\tilde{\underline{x}}$ of the true disc is an element of $\Delta_2 = \Theta_1 \cap \Theta_2$. Fig. 5.5c and Fig. 5.5d illustrate the solution set Δ_4 for four measurements. The discs $\mathbf{K}(\tilde{\underline{x}})$, $\mathbf{K}\left(\underline{x}_4^{(1)}\right)$ and $\mathbf{K}\left(\underline{x}_4^{(2)}\right)$ are examples for feasible circular discs, where $\mathbf{K}\left(\underline{x}_4^{(1)}\right)$ is the smallest enclosing circle for the given measurements and $\mathbf{K}(\tilde{\underline{x}}_4)$ is the true disc. Fig. 5.5d shows the solution set Δ_4 and the parameters of the example discs, i.e., $\underline{x}_4^{(1)}$ and $\underline{x}_4^{(2)}$.

Remark 5.7. For a stationary circle and noise-free measurements the following holds: The apex of $\Delta_{k,l}$ converges to the parameter vector of the smallest enclosing circle of the object in case the measurements cover the entire area of the object with an increasing number of measurements.

Because the exact recursive computation of the solution set is infeasible, we introduce a novel outer-bounding technique that exploits that the intersection of a cone with a hyperboloid can be outer-bounded by a hyperboloid

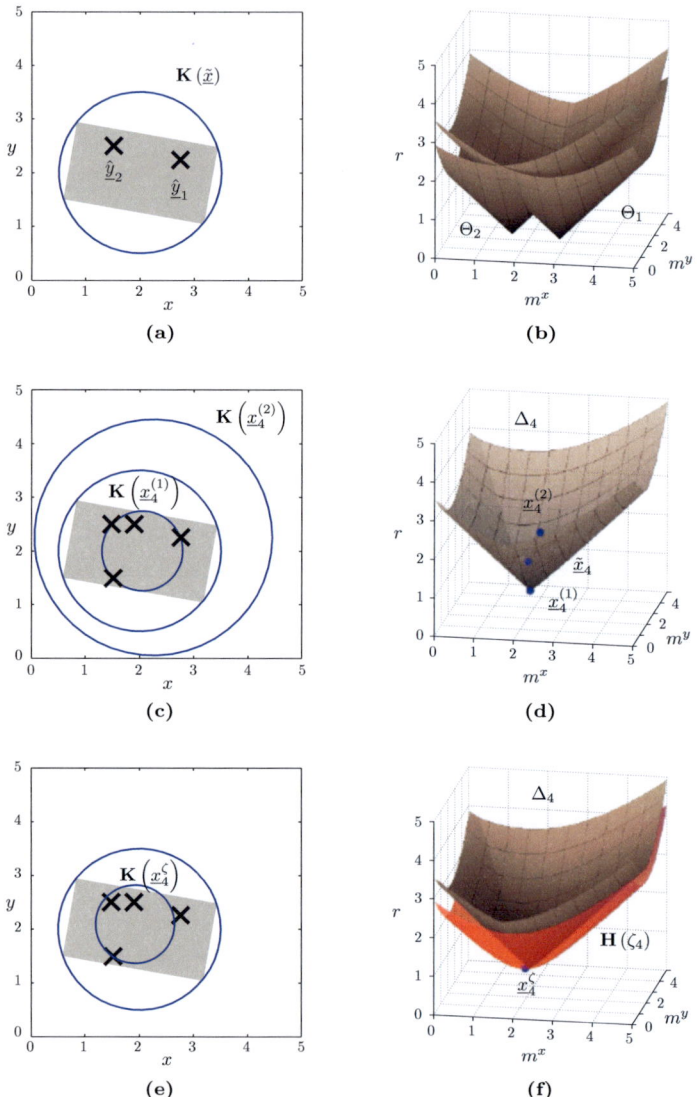

Figure 5.5: Set-theoretic estimation of the parameters of a circular disc.

again. With the help of this approximation technique, it is possible recursively outer-bound the exact solution set with hyperboloids such that the convergence property in Remark 5.7 still holds.

Definition 5.2 (Hyperboloid of Revolution). The upper sheet of a two-sheeted circular hyperboloid of revolution is given by

$$\mathbf{H}\left(\zeta^x, \zeta^y, \zeta^z, \zeta^a\right) := \{[m^x, m^y, r]^T \in \mathbb{R}^3 \mid r \geq \zeta^z$$
$$\text{and } (m^x - \zeta^x)^2 + (m^y - \zeta^y)^2 + (\zeta^a)^2 \leq (r - \zeta^z)^2\} \quad (5.11)$$

with $\zeta^x, \zeta^y, \zeta^z \in \mathbb{R}$ and $\zeta^a \in \mathbb{R}^+$.

Remark 5.8. The hyperboloid (5.11) has the following properties (see Figure 5.6a for an illustration):

- The focus is $F = [\zeta^x, \zeta^y, \zeta^z]^T$.

- The apex is located at $A = [\zeta^x, \zeta^y, \zeta^a + \zeta^z]^T$.

- The cone angle is orthogonal.

- It is oriented along the r–axis.

Definition 5.3 (Cone). The set $\mathbf{C}\left(\zeta^x, \zeta^y\right) := \mathbf{H}\left(\zeta^x, \zeta^y, 0, 0\right)$ is a cone oriented along the r–axis whose apex lies on the $m^x m^y$–plane.

An *SSI filter* can be constructed by representing the uncertainty about the center and radius as a random hyperboloid $\mathbf{H}\left(\underline{\zeta}_{k,l-1}\right)$. When a new measurement becomes available, the random hyperboloid is intersected with the corresponding random solution set $\mathbf{C}\left(\hat{\underline{y}}_{k,l} - \underline{v}_{k,l}\right)$. This intersection is outer-bounded by a hyperboloid $\mathbf{H}\left(\underline{\zeta}_{k,l}\right)$ again, i.e.,

$$\mathbf{H}\left(\underline{\zeta}_{k,l}\right) \supseteq \mathbf{H}\left(\underline{\zeta}_{k,l-1}\right) \cap \mathbf{C}\left(\underline{z}_{k,l}\right) \ ,$$

where $\underline{z}_{k,l} := \hat{\underline{y}}_{k,l} - \underline{v}_{k,l}$. The parameter vector $\underline{\zeta}_{k,l}$ can be calculated from $\underline{\zeta}_{k,l-1}$ and $\underline{z}_{k,l}$ as shown in the following theorem and the subsequent definition (see also Fig. 5.6 and Fig. 5.7).

Theorem 5.2. *Suppose we are given a cone* $\mathbf{C}\left(\underline{z}\right)$ *with* $\underline{z} = \left[z^x, z^y\right]^T$ *and a hyperboloid* $\mathbf{H}\left(\zeta\right)$ *with* $\underline{\zeta} = \left[\zeta^x, \zeta^y, \zeta^z, \zeta^a\right]^T$. *The distance between the vectors* $\left[z^x, z^y\right]^T$ *and* $\left[\zeta^x, \zeta^y\right]^T$ *is denoted as*

$$d := \sqrt{(z^x - \zeta^x)^2 + (z^y - \zeta^y)^2} \ .$$

If the condition

$$d > \zeta^z + \zeta^a \tag{5.12}$$

holds, the hyperboloid $\mathbf{H}\left(\zeta_{Bound}\right)$ *with*

$$\begin{bmatrix} \zeta^x_{Bound} \\ \zeta^y_{Bound} \end{bmatrix} = \begin{bmatrix} z^x \\ z^y \end{bmatrix} + \frac{1}{d} \cdot (m z_{apex} + c) \cdot \left(\begin{bmatrix} \zeta^x \\ \zeta^y \end{bmatrix} - \begin{bmatrix} z^x \\ z^y \end{bmatrix} \right) \ , \tag{5.13}$$

$$\zeta^z_{Bound} = m^2 z_{apex} + mc \ , \tag{5.14}$$

$$\zeta^a_{Bound} = z_{apex} - \zeta^z_{Bound} \ , \tag{5.15}$$

where $m = \frac{\zeta^z}{d}$, $c = d + \frac{(\zeta^a)^2 - (\zeta^z)^2}{2d}$ *and*

$$z_{apex} = \begin{cases} -\frac{c}{m-1} & \text{if } m^2 - 1 \neq 0 \\ -\frac{c}{2m} & \text{if } m^2 - 1 = 0 \end{cases}$$

fulfills the following properties:

1. *The apex of* $\mathbf{C}\left(\underline{z}\right) \cap \mathbf{H}\left(\zeta\right)$ *is* $\left[\zeta^x_{Bound}, \zeta^y_{Bound}, z_{apex}\right]^T$, *where* $z_{apex} = \zeta^a_{Bound} + \zeta^z_{Bound}$.

2. $\partial \mathbf{C}\left(\underline{z}\right) \cap \partial \mathbf{H}\left(\zeta\right)$ *is a hyperbola that lies in a plane* E *with normal vector* $\left[\zeta^x - z^x, \zeta^y - z^y, \zeta^z_{Bound}\right]^T$ *and position vector* $\left[\zeta^x_{Bound}, \zeta^y_{Bound}, z_{apex}\right]^T$.

3. $E \cap \partial \mathbf{H}\left(\zeta_{Bound}\right) = \partial \mathbf{C}\left(\underline{z}\right) \cap \partial \mathbf{H}\left(\zeta\right)$.

4. $\mathbf{C}\left(\underline{z}\right) \cap \mathbf{H}\left(\zeta\right) \subseteq \mathbf{H}\left(\zeta_{Bound}\right)$.

Proof. Can be proven with basic algebraic rules. $\qquad\qquad\square$

Remark 5.9. Condition 5.12 means that the projection of the apex of $\mathbf{C}\left(\underline{z}\right) \cap \mathbf{H}\left(\zeta\right)$ onto the $m^x m^y$-plane lies on the segment from $\left[z^x, z^y, 0\right]^T$ to $\left[\zeta^x, \zeta^y, 0\right]^T$.

Definition 5.4. The function $\mathcal{G}_2 : \mathbb{R}^7 \to \mathbb{R}^4$ is defined as follows

$$\mathcal{G}_2(\underline{z}, \zeta) = \begin{cases} \mathcal{G}_2^*(\underline{z}, \zeta) & \text{if (5.12) holds} \\ \left[\underline{z}^T, 0\right]^T & \text{if } \zeta^z + \zeta^a < 0 \\ \zeta & \text{otherwise} \end{cases},$$

where $\mathcal{G}_2^* : \mathbb{R}^7 \to \mathbb{R}^4$ denotes the function specified by (5.13) - (5.15) that maps $\underline{z}$ and ζ to ζ_{Bound}.

Based on the above definition, an *SSI filter* for the center and radius of a circular disc can be constructed.

Statistical and Set-Theoretic Information (SSI) filter 2

- **Solution Set**
  $\mathbf{H}\left(\underline{\zeta}_{k,l}\right)$ with $\underline{\zeta}_{k,l} = \left[\zeta_{k,l}^x, \zeta_{k,l}^y, \zeta_{k,l}^z, \zeta_{k,l}^a\right]^T$
  and $\underline{\zeta}_{k,l} \sim \mathcal{N}(\zeta - \hat{\underline{\zeta}}_{k,l}, \Sigma_{k,l}^\zeta)$.

- **Measurement Solution Set**
  $\mathbf{C}\left(\underline{z}_{k,l}\right)$ with $\underline{z}_{k,l} := \hat{\underline{y}}_{k,l} - \underline{w}_{k,l}$ and $\underline{z}_{k,l} \sim \mathcal{N}(z - \hat{\underline{y}}_{k,l}, \mathbf{C}_{k,l}^z)$.

- **Measurement Update**

  $$\underline{\zeta}_{k,l} = \mathcal{G}_2(\underline{z}_{k,l}, \underline{\zeta}_{k,l-1}) \tag{5.16}$$

- **Time Update**
  A random walk model results in the following system equation for the parameters of the solution set

  $$\underline{\zeta}_{k+1,0} = \underline{\zeta}_{k,n_k} + \begin{bmatrix} \mathbf{I}_2 \\ \mathbf{0}_2 \end{bmatrix} (\hat{\underline{u}}_k + \underline{w}_k) + b_k \underline{e}_4^{(4)}, \tag{5.17}$$

  where $\underline{e}_4^{(4)} = [0,0,0,1]^T$ and b_k is a scalar that accounts for an unknown-but-bounded error in addition to the stochastic noise term $\underline{w}_k$.

Remark 5.10. When there is neither measurement nor system noise, the apex of $\mathbf{H}\left(\underline{\zeta}_{k,l}\right)$, namely $\zeta_{k,l}^z + \zeta_{k,l}^a$, converges to the true parameters if for each time step, the future measurements cover the entire extended object. This convergence property is something special about SSI 2 as it says that

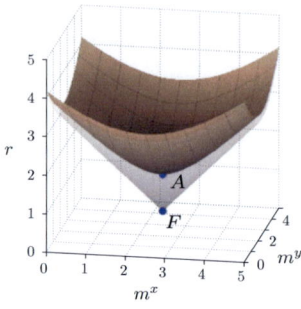

(a) Hyperboloid with apex A and focus F.

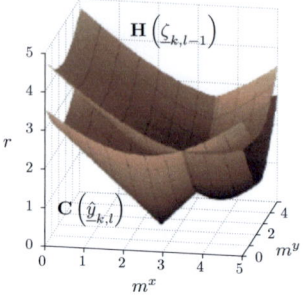

(b) Cone $\mathbf{C}(2.5, 2)$ and a hyperboloid $\mathbf{H}(4, 2, 1)$.

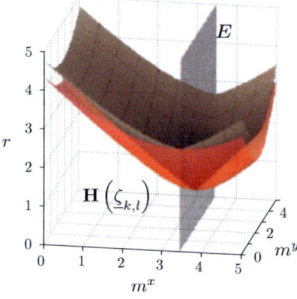

(c) Outer bound of the intersection.

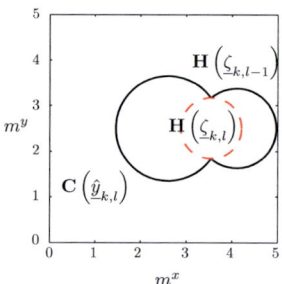

(d) $m^x m^y$-plane for radius $r = 1.8$.

Figure 5.6: Outer-bounding the intersection of a cone and a hyperboloid (with $\zeta_{k,l-1}^z = 0$) with a hyperboloid.

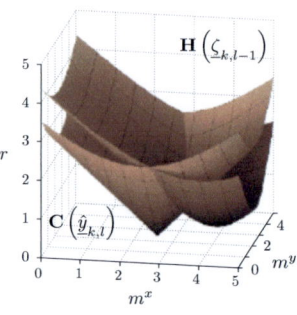

(a) Cone $\mathbf{C}\left(\begin{bmatrix} 2.5, 2 \end{bmatrix}\right)$ and hyperboloid $\mathbf{H}\left([4, 2, -0.5, 1]^T\right)$.

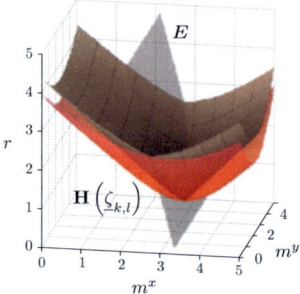

(b) Outer bound of the intersection.

Figure 5.7: Outer-bounding the intersection of a cone and a hyperboloid (when $\zeta_{k,l}^z > 0$) with a hyperboloid.

the apex converges to the parameters of the smallest enclosing circle of the object, regardless of the true shape. In this manner, SSI 2 encompasses an algorithm for computing the smallest enclosing shape based on recursively arriving point measurements.

Remark 5.11. If there is measurement noise, but the support of the measurement noise $\underline{v}_{k,l}$ is bounded, SSI 2 converges to the true circular disc plus the bounded support. Otherwise, SSI 2 does not converge to a fixed value in general. The reason is that both the stochastic and set-valued uncertainties cannot vanish as the intersection of two measurement solution sets is always unbounded.

Example 5.3. Fig. 5.5f depicts an example of the resulting hyperboloid $\mathbf{H}\left(\underline{\zeta}_4\right)$ after four measurements. The circle specified by its apex $\underline{x}_4^\zeta$ does not enclose all received measurements (see Fig. 5.5e), because $\mathbf{H}\left(\zeta_4\right)$ is a conservative approximation of the true solution set, but $\tilde{\zeta}$ is an element of $\mathbf{H}\left(\zeta_4\right)$.

Remark 5.12. For given radius, SSI 2 equals to the SSI 1 for circular discs with known radius. The random solution set $\Delta_{k,l}^r$ for the center given the radius results from the random hyperboloid $\mathbf{H}\left(\underline{\zeta}_{k,l}\right)$, i.e.,

$$\Delta_{k,l}^r = \mathbf{K}\left(\zeta_{k,l}^x, \zeta_{k,l}^y, \sqrt{(\tilde{r}_k)^2 - (\zeta_{k,l}^a + \zeta_{k,l}^z)^2}\right)$$

given that $\zeta_{k,l}^a + \zeta_{k,l}^z \le \tilde{r}_k$.

Estimating the Radius As already mentioned earlier, the radius of circular disc cannot be estimated if the extent is modeled as an unknown-but-bounded error. This problem can be by-passed when a further measured quantity allows for inferring the shape parameters.

A reasonable assumption is that the number of received measurements at a particular time step depends on the object's size. For example, [Koc08] assumes that the number of received measurements is Poisson distributed with mean that is proportional to the area of the shape. In the following, we assume that the number of measurements depends on the actual radius of the extended object.

In order to enhance SSI 2 for estimating the radius, an additional state variable $r_{k,l}$ for the information inferred from the number of measurements

n_k is maintained, i.e., the state vector is now $\left[\underline{\varsigma}_{k,l}^T, r_{k,l}\right]^T$. The random variable $r_{k,l}$ can be updated with Bayes' rule.

Statistical and Set-Theoretic Information (SSI) filter 3

- **State vector**
  $\left[(\underline{\varsigma}_{k,l})^T, r_{k,l}\right]^T$ with $\underline{\varsigma}_{k,l} = \left[\varsigma_{k,l}^x, \varsigma_{k,l}^y, \varsigma_{k,l}^z, \varsigma_{k,l}^a\right]^T$ and the (implicit) state constraint $\varsigma_{k,l}^a + \varsigma_{k,l}^z > r_{k,l}$

- **Solution Set**
  $\mathbf{H}\left(\underline{\varsigma}_{k,l}\right)$

- **Measurement Solution Set**
  $\mathbf{C}\left(\underline{z}_{k,l}\right)$ with $\underline{z}_{k,l} := \hat{\underline{y}}_{k,l} - \underline{w}_{k,l}$ and $\underline{z}_{k,l} \sim \mathcal{N}(z - \hat{\underline{y}}_{k,l}, \Sigma_{k,l}^v)$.

- **Measurement Update**

  – Update of the solution set for the center

  $$\underline{\varsigma}_{k,l} = \mathcal{G}_2(\underline{z}_{k,l}, \underline{\varsigma}_{k,l-1}) \qquad (5.18)$$

  – Measurement equation for radius update

  $$n_k = \underline{r}_k + \underline{v}_k^r \qquad (5.19)$$

  – State constraint enforcing

  $$\varsigma_{k,l}^a + \varsigma_{k,l}^z > r_{k,l} \qquad (5.20)$$

- **Time Update**
  A random walk model results in the following system equation for the parameters of the solution set

$$\begin{bmatrix} \underline{\varsigma}_{k+1,0} \\ r_{k+1,0} \end{bmatrix} = \begin{bmatrix} \underline{\varsigma}_{k,n_k} \\ r_{k,n_k} \end{bmatrix} + \begin{bmatrix} \mathbf{I}_2 \\ \mathbf{0}_4 \end{bmatrix} (\hat{u}_k^m + \underline{w}_k^m) + (\underline{e}_3^{(5)} + \underline{e}_5^{(5)}) \cdot (\hat{u}_k^r + \underline{w}_k^r) - \underline{e}_3^{(5)} b_k \ ,$$

$$(5.21)$$

where $(\underline{e}_3^{(5)} + \underline{e}_5^{(5)}) \cdot (\hat{u}_k^r + \underline{w}_k^r)$ captures the stochastic system noise for the radius, b_k is a scalar accounting for an unknown-but-bounded error on the center, and $\underline{e}_3^{(5)} = [0, 0, 1, 0, 0]^T$ and $\underline{e}_5^{(5)} = [0, 0, 0, 0, 1]^T$ are unit vectors.

Herein, the mapping (5.18) is the same as in SSI 2. The radius estimate is updated based on (5.19), which relates the radius with the number of measurements. As we aim at a Gaussian filter, it is suitable to consider n_k as a continuous random variable (although it is in fact discrete). If the measurement noise $\underline{v}_k^r$ is additionally Gaussian, (5.19) essentially says that the number of measurements is Gaussian distributed with mean propoprtional to the radius. The update (5.18) can be performed with the Kalman filter formulas.

Equation (5.20) is a constraint on the state vector. If this constraint is not fulfilled, the solution set for the center $\mathbf{H}\left(\underline{\varsigma}_{k,l}\right)$ given the radius is empty. This constraint has to be enforced in order to reduce the uncertainty of the state estimate. The enforcement of a linear inequality constraint as (5.20) can be performed by probability density truncation. The truncated density can be re-approximated with a Gaussian distribution by calculating the first two moments of the truncated density as described in [Sim06].

Remark 5.13. Note that SSI 3 turns into SSI 1 for given $\boldsymbol{r}_k$.

5.6   Evaluation

The benefits of the set-theoretic extent model are demonstrated by means of tracking the aircaft-shaped extended object in Fig. 5.8. There are three possible sets of measurement sources on the extended object that are alternated randomly between two time steps. The measurement sources itself are drawn uniformly from the particular set of measurement sources. Such a scenario could be caused by specific surface characteristics so that only particular parts of the objects give rise to reflections.

Scenario 1: Known Radius

In the first setting, we compare SSI 1 for known radius $\tilde{r}_k = 2.3$ with the plain stochastic estimator that employs a spatial distribution model for the extent as already described in Remark 5.2 for the motivating example.

The statistics of the measurement noise $\Sigma_{k,l}^v$ are the same for all measurements. However, we consider scenarios with three different noise levels

- Scenario 1a: $\Sigma_{k,l}^v = \mathrm{diag}([0.05, 0.05]^T)$ (small noise),

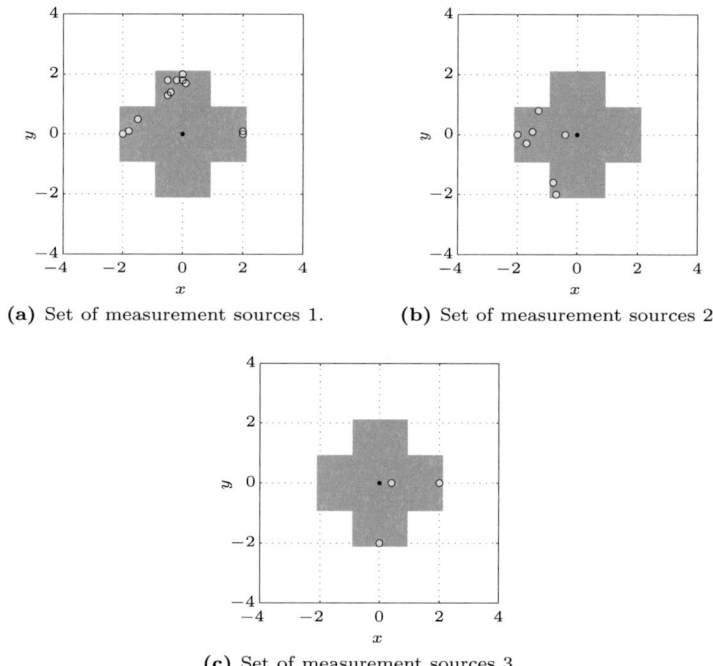

(a) Set of measurement sources 1.   **(b)** Set of measurement sources 2.

(c) Set of measurement sources 3.

Figure 5.8: Three sets of possible measurement sources for the considered object.

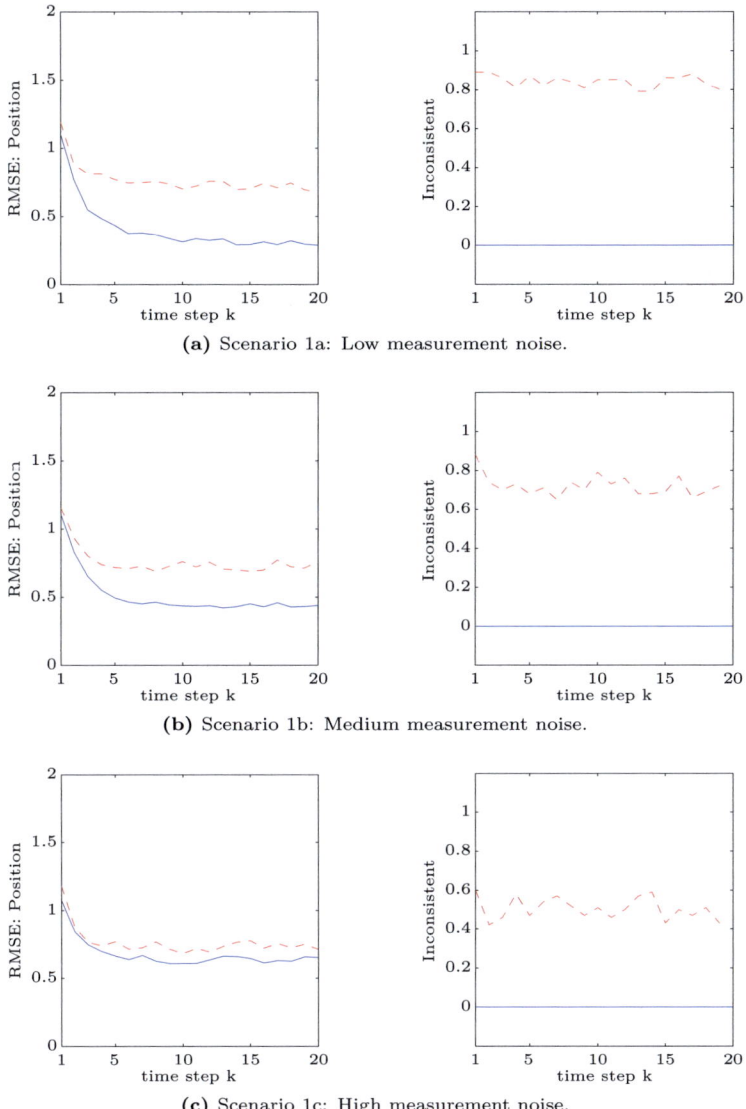

(a) Scenario 1a: Low measurement noise.

(b) Scenario 1b: Medium measurement noise.

(c) Scenario 1c: High measurement noise.

Figure 5.9: Simulation results: RMSE for position and percentage of inconsistent estimation results (for 100 runs).

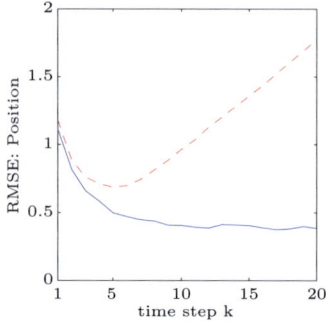 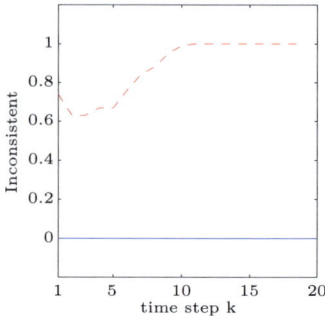

Figure 5.10: Scenario 1d: Low system noise.

- Scenario 1b: $\Sigma_{k,l}^v = \mathrm{diag}([0.1, 0.1]^T)$ (medium noise),

- Scenario 1c: $\Sigma_{k,l}^v = \mathrm{diag}([0.2, 0.2]^T)$ (large noise), and

- Scenario 1d: $\Sigma_{k,l}^v = \mathrm{diag}([0.1, 0.1]^T)$ (medium noise).

The number of measurements n_k is drawn from a Gaussian distribution with mean $8 \cdot \tilde{r}_k$ and covariance 3, where the result is rounded and negative values are rejected.

SSI 1 specifies the temporal evolution of the position by a random walk model (2.4) with input $\hat{\underline{u}}_k = [7, 0, 0]^T$ and $b_k = 0.005$, and additive noise as follows.

- Scenario 1a - c: $\mathbf{C}_k^v = \mathrm{diag}([0.02, 0.06]^T)$,

- Scenario 1d: $\mathbf{C}_k^v = \mathrm{diag}([0.02, 0.002]^T)$.

The same motion model is used for the stochastic estimator, where the unknown-but-bounded error is converted to a Gaussian distribution in the same way as in the measurement equation in Remark 5.1.

Discussion   Fig. 5.9 shows the *Root-Mean-Square Error* (*RMSE*) for the position estimates for both SSI 1 and the stochastic estimator. It can be seen that SSI 1 yields significantly better estimation results than the

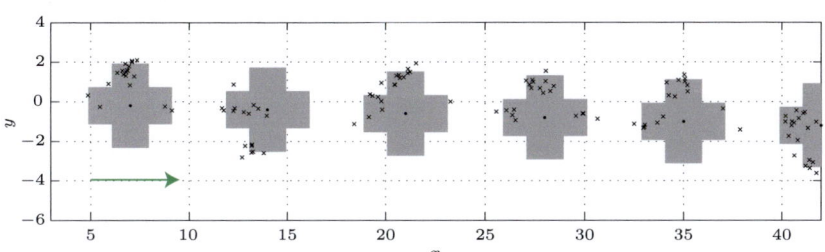

(a) Extended object plus the received measurements plotted over several time steps.

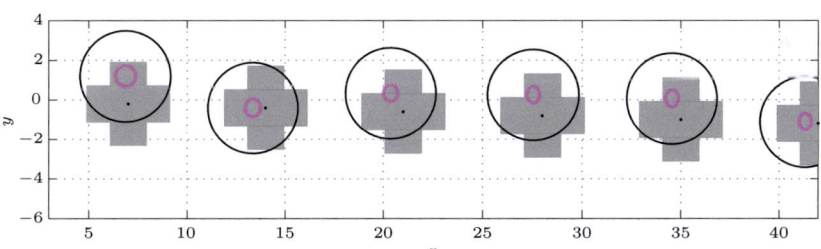

(b) Estimation results of the Bayesian estimator. The 95% confidence ellipse for the center is plotted in magenta.

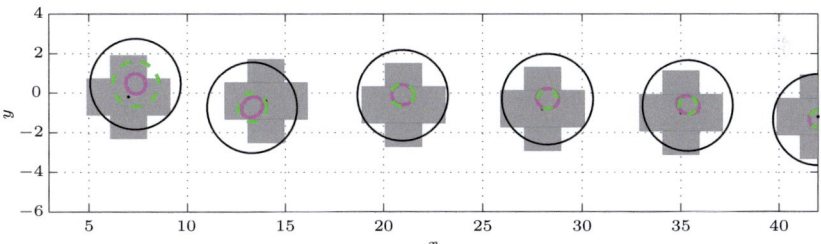

(c) Estimation result of the *SSI filter* using a set-theoretic extent model. The 95% confidence ellipse for the center is plotted in magenta and the set-theoretic error is indicated by a dashed green circle.

Figure 5.11: Scenario 1: Tracking an extended object characterized by biased measurement sources.

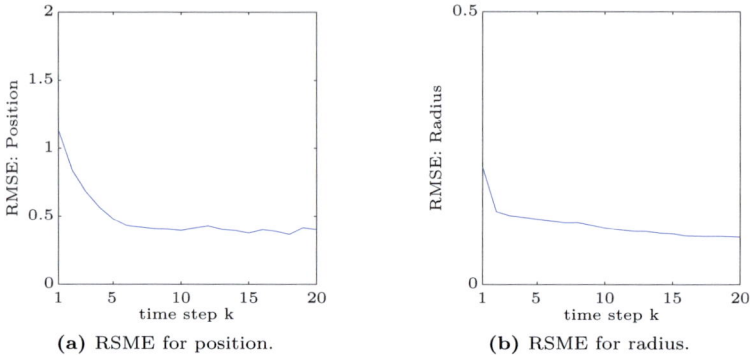

(a) RSME for position.   **(b)** RSME for radius.

Figure 5.12: Estimation results when the radius is unknown.

stochastic estimator as the locations of the measurement sources are biased. However, with an increasing measurement noise, the differences vanish. Furthermore, Fig. 5.9 depicts the percent of inconsistent estimates. A point estimate is considered as *consistent* if the confidence region contains the true value (see also Definition 5.1). The stochastic estimator frequently gives inconsistent results. This is a consequence of the high number of measurements per time instant. In this case, the estimates approach the mean of the measurement sources and the stochastic uncertainties tend to vanish. However, SSI 1 nearly always gives consistent results. Even the large number of measurements does not result in a decrease of the set-valued uncertainties as they do not cover the entire surface. In Scenario 1d, the system noise is slightly decreased. As a consequence, the stochastic estimator even looses the track, which is very disadvantageous.

It is very important to note that increasing the system noise does not solve the inconsistency problem with the stochastic estimator. With an increasing number of measurements per time step n_k, the stochastic estimator is bound to become inconsistent in case of biased measurement sources.

Scenario 2: Unknown Radius

The second setting is essentially the same as Scenario 1b except that the radius is also unknown and part of the estimation problem. The purpose of the setting is to show that SSI 2 is capable of estimating the radius in addition to the center. As only the center of the circle is modeled as a set-theoretic error and the radius is a probabilistic quantity, a comparison with a purely stochastic estimator is not necessary (they would be equal). A priori the radius is assumed to be 3.3 with a variance of 0.5. Fig. 5.12 depicts the *RMSE* for the radius. As the radius does not change over time, the estimated radius approaches the true one after a couple of time steps.

5.7 Conclusions

Typically, extended object tracking methods assume that each measurement source is an independent random draw of a particular probability distribution whose mass is concentrated on the object's extent. However, this is in fact an ad-hoc assumption as in many applications, there is no (statistical) information about the measurement sources available. For example, the locations of the measurement sources may be significantly influenced by the properties of the object's surface.

In this chapter, we took an extreme position: No assumptions about the location of the measurement sources on the extended object have been imposed, i.e., the extent was modeled as an unknown-but-bounded error. For this kind of model, we derived novel *SSI Filters* for tracking circular discs. In particular, we employed random hyperboloids in order to express the uncertainty about the center and radius of the circular disc, and we derived outer-bounding techniques for a recursive processing of measurements. A direct consequence of the set-theoretic extent model is that it is in general not possible to estimate the parameters of the shape based on point measurements. Hence, this approach is in general only suitable if the shape is given or can be inferred from other information sources. In particular, we suggested to estimate the radius of the circular disc based on the number of received measurements.

With the help of simulations, it was demonstrated that systematic errors in the measurement sources may lead to inconsistent estimation results when using a pure stochastic state estimator, where a point estimate is called inconsistent if the associated confidence region does not contain the true value. The novel *SSI filter*, however, is capable of providing consistent and precise estimation results. Nevertheless, the simulations also demonstrated that the systematic error becomes negligible when the measurement noise increases.

In summary, the proposed *SSI Filters* based on the set-theoretic extent models are suitable when

- the locations of the measurement sources are dominated by systematic errors,

- the measurement noise is rather low compared to the extent, and

- the shape parameters are known or can be estimated based on additional information.

Nonetheless, it is important to note that even under the above conditions, the *SSI filter* does not always provide better estimation results than the pure Bayesian approach regarding the *Root-Mean-Square Error*. This results from the fact that if the modeling assumptions of a Bayesian estimator are fulfilled, the Bayesian estimator (if calculated exactly) provides the *Minimum Mean Squared Error (MMSE)* estimate. In general, the *SSI filter* is capable of providing consistent estimation results.

So far, we focused on circular discs. Of course, the presented approach is also suitable for other shapes such ellipsoids, rectangles or polygons. However, the set-bounding techniques have to be tailored to the particular shape. Hence, the challenge is to find a proper conservative approximation of solution sets. For this purpose, approximation techniques from set-membership estimation such as bounding ellipsoids and polytopes can be exploited. An *SSI filter* for rectangular shapes has been developed in [10].

CHAPTER 6

Moving Object Tracking based on RGBD Data – Experiments

This chapter is about an experimental setup for evaluating extended object tracking algorithms. The task is to track an unknown, ground moving object from a bird's eye view with the help of a camera that provides both RGB and depth images. We focus on a single moving object and assume that everything else in the scene is stationary. The surveillance area is a table on which a toy train travels on tracks. A Microsoft® Kinect™ sensor captures the RGB and depth images (see Fig. 6.1).

In order to discriminate the object from the background, moving points are detected in the RGB and depth image sequences (see Fig. 6.2c). By this means, a set of noisy measurements stemming from the object's surface is received for each frame. Due to the sensor noise, the measurements are noisy and do not completely cover the surface of the object. Hence, the estimation of the shape and kinematic parameters based on the extracted measurements is a typical *extended object tracking problem* as introduced in Chapter 2. We employ star-convex *RHMs* as described in Chapter 4.3 for this purpose and highlight the advantages compared to a variant of active contour models.

The experimental setting mainly serves as a test-environment for extended object tracking methods because the characteristics of the measurements resemble larger sensors such as *Ground Moving Target Indicator (GMTI)* sensors [KKU06]. Additionally, the overall tracking system has several advantages compared to standard tracking methods. For example, the combination of RGB and depth data is especially advantageous for moving point detection, because moving objects almost always differ from the background in depth and usually in color.

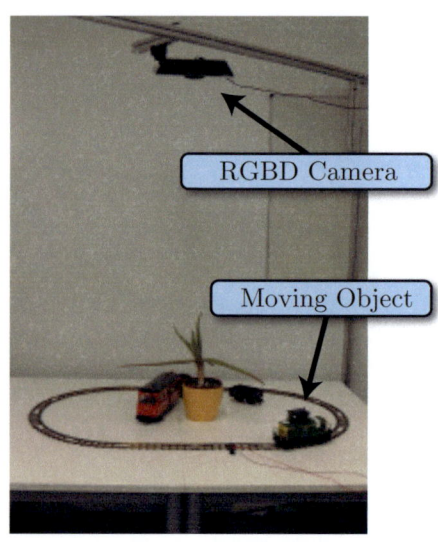

Figure 6.1: Setup: Moving object, i.e., a toy train, observed with an RGBD camera from a bird's eye view [1], [5].

We pursue the philosophy of performing as little preprocessing of the raw data, i.e., the RGB and depth images, as required. The signal processing algorithm that detects moving points utilizes rather basic, model-free image processing techniques, e.g., no background model is learned. All modeling knowledge is put into the tracking algorithm. Extended object tracking algorithms based on *RHMs* deal with the extracted data in a systematic and mathematically well-grounded manner using a probabilistic model for a *single* point measurement. It is essential to note that no prior information about the shape or color are required. The extended object tracking algorithm estimates the object shape from scratch and shape changes are tracked based on a probabilistic system model for the shape (see Section 2.1.3).

Remark 6.1. This chapter is a revised version of [5], where the experimental setting, the signal processing algorithm, and some preliminary experiments have been presented. In [30], the real-time implementation of the tracking system has been developed. A brief overview of the system has been given in [1].

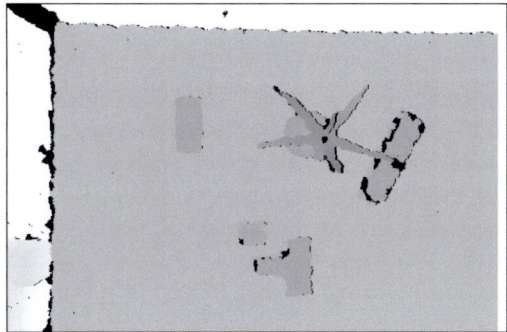

(a) RGB image.

(b) Depth image.

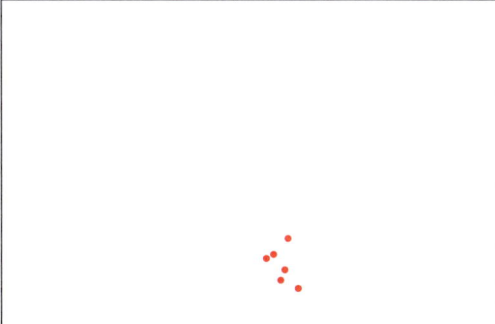

(c) Extracted measurements that serve as input for the tracking algorithm. Measurements are only received from the train as everything else is stationary.

Figure 6.2: Signal processing: RGB and depth images supplied by the Kinect and the extracted measurements.

6.1   Related Work

Recent overviews of RGB-only object tracking algorithms can be found in [YSZ+11, YJS06]. A standard technique for visual shape tracking is the so-called active contour model. An active contour model determines the object contour by minimizing an energy function [KWT88, CKS95] that consists of an internal force for regularization and an external force for pushing the contour to image features. A further popular class of visual tracking methods are so-called kernel methods [CRM03]. For example, the object may be represented with a color histogram in an ellipsoidal region and the location of the region is tracked by maximizing a similarity function using the mean-shift algorithm. Both active contours and kernel methods are fundamentally different from extended object tracking methods that are based on the assumption of independently generated point measurements.

Moving object detection (see [YJS06, YP05]) is a widely-used technique for background subtraction in visual object tracking. The most simple approach for detecting moving points is to calculate frame differences. More elaborate techniques such as [GBCR00] learn a background model for the color.

In the context of remote sensing, *Moving Target Indication (MTI)* [KKU06] denotes special mode of a radar for discriminating an object from the background by means of the Doppler effect, e.g., a *Ground MTI (GMTI)* radar [KKU06] detects ground moving objects. A related concept to radar-based MTI is *Visual MTI (VMTI)* [JBR06], which detects ground moving objects from a bird's eye view based on visual data.

6.2   Signal Processing - Moving Point Detection

In this section, the signal processing algorithm for detecting moving points is explained. The RGBD sensor captures a sequence of

- RGB images $I_k(x, y) \in \mathbb{R}^3 \cup \perp$, and

- depth images $D_k(x, y) \in \mathbb{R} \cup \perp$,

where k denotes the discrete time index, and $[x, y]^T \in \{1, 2, \ldots, n_u\} \times \{1, 2, \ldots, n_v\}$ is a pixel coordinate in an image with resolution $n_u \times n_v$. The symbol $\perp$ is used in the following to tag a pixel as "ground".

For each frame at time k, the task of the signal processing algorithm is to compute a set of pixel coordinates that are classified as "moving"

$$\hat{\mathcal{Y}}_k = \{\underline{\hat{y}}_{k,1}, \ldots, \underline{\hat{y}}_{k,n_k}\} \; ,$$

where $\underline{\hat{y}}_{k,l} \in \{1, 2, \ldots, n_u\} \times \{1, 2, \ldots, n_v\}$ for $l \in \{1, \ldots, n_k\}$. The pixel coordinates in $\hat{\mathcal{Y}}_k$ are then taken as the measurements for the extended object tracking algorithm (see the next section). The signal processing algorithm involves three consecutive steps, where the actual moving point detection is performed in the second step.

Step 1: Remove Ground A realistic assumption is that a moving object differs in height from the ground, i.e., the table. As the distance from the table to the sensor is known, pixels that belong to ground and therefore do not move can be determined. For this purpose, each pixel that exceeds a user-defined distance t_{depth} from the sensor is tagged as "ground", where t_{depth} is the distance from the sensor to the table. The RGB and depth images after detecting pixels from the ground are

$$D_k^t(x, y) := \begin{cases} D_k(x, y) & \text{if } D_k(x, y) < t_{\text{depth}} \\ \perp & \text{otherwise} \end{cases} ,$$

and

$$I_k^t(x, y) := \begin{cases} I_k(x, y) & \text{if } D_k(x, y) < t_{\text{depth}} \\ \perp & \text{otherwise} \end{cases} ,$$

where $\perp$ tags a pixel as "ground". Ground pixels are ignored in the detection process as they are always stationary.

Step 2: Detect Moving Points In the second step, moving points are detected by utilizing a standard optical flow algorithm, i.e., the Horn-Schunck method [HS81]. The optical flow algorithm is applied to the RGB image sequence $I_k^t(x, y)$ and depth image sequence $D_k^t(x, y)$ *separately*. The velocity of pixels marked as "ground" is set to zero. As a result, two velocity fields are obtained, i.e., $I_k^v(x, y) \in \mathbb{R}^2$ for the RGB images and $D_k^v(x, y) \in \mathbb{R}^2$ for the depth images. A particular pixel is classified as "moving" if the magnitude of its velocity vector exceeds a user-defined threshold either in the RGB *or* depth image sequence, i.e.,

$$\hat{\mathcal{Y}}_k^{\text{mov}} := \left\{ [x, y]^T \;\middle|\; \|I_k^v(x, y)\| > t_{\text{vel}} \text{ or } \|D_k^v(x, y)\| > t_{\text{vel}} \right\} . \quad (6.1)$$

Step 3: Reduce Clutter Owing to the structured light approach of the sensor for measuring depth, contours of rather small objects heavily jitter in the depth images. In the specific considered setting, this is a serious problem as boundary points of non-moving objects are frequently classified as "moving". Our solution to this problem is to detect these clustered detections in $\hat{\mathcal{Y}}_k^{\text{mov}}$ and dismiss them. A detected moving point is classified as clutter, i.e., it is part of a spurious cluster detection, if a particular number n_{edge} of neighbors is moving. Herein, a neighbor is a point whose distance to the considered point is at most r_{edge}.

$$\hat{\mathcal{Y}}_k := \left\{ [x, y]^T \in \hat{\mathcal{Y}}_k^{\text{mov}} \;\middle|\; \# \left\{ \hat{\mathcal{Y}}_k^{\text{mov}} \cap \mathbf{K} \left([x, y]^T, r_{\text{edge}} \right) \right\} \le n_{\text{edge}} \right\} ,$$
$$(6.2)$$

where $\mathbf{K} \left([x, y]^T, r_{\text{edge}} \right)$ denotes all points whose distance to $[x, y]^T$ is less than r_{edge}. As a consequence, false detections on the edges are removed and only moving points from the object are obtained.

The resulting set of pixels $\hat{\mathcal{Y}}_k$ serves as input for the extended object tracking algorithm. Fig. 6.2c depicts the extracted measurements $\hat{\mathcal{Y}}_k$ for an example frame.

Parameters to Adjust

The signal processing algorithm involves a couple of parameters that have to be adjusted. From a practicable point of view, it is simple to find them and most importantly they are not object dependent, i.e., the same parameters can be used for all objects.

In step 1: t_{depth} The distance from the sensor to the table t_{depth} can either be measured directly manually or an automatic procedure can be implemented easily.

In step 2: t_{vel} The threshold on the velocity t_{vel} should be chosen such that a stationary object does not give rise to measurements, but a moving object does. In general, a small value of the velocity threshold leads to an increasing number of detected measurements on the object. However, a small velocity threshold usually also increases the false measurement rate, i.e., the number of pixels that are wrongly classified as "moving". If the velocity threshold is too large, only few measurements from the object are detected, which may render the shape tracking challenging. Hence, when determining t_{vel}, it is essential to find a suitable trade-off between object generated measurements and false measurements. Note that in the signal processing of radar data, there is a parameter called detection threshold, which has essentially the same meaning as t_{vel}. If the signal strength of a received radar echo exceeds the detection threshold, it is considered as target generated. Due to this analogy, the behavior of the obtained measurements is very similar to the measurements received from a radar device.

In step 3: n_{edge} **and** r_{edge} The threshold for detecting edges n_{edge} and r_{edge} can be found by visual inspection. They have to be chosen in a manner that edges of stationary objects disappear. In general, these parameters could be calculated automatically based on the specification of the Kinect sensor. They mainly depend on the distance from the object to sensor.

6.3   Shape Tracking with Random Hypersurface Models

For each time step k, the signal processing algorithm supplies a varying number of noisy measurements $\hat{\mathcal{Y}}_k$ from the object surface (see Fig. 6.2c). Owing to the low sensor resolution and the small extent of the object, only a limited number of measurements that do not completely cover the object's surface is detected. Hence, the characteristics of the measurements suggest an extended object tracking algorithm. In particular, the *UKF* implementation for star-convex RHMs *UKF-SC-RHM* as described in Section 4.3 is used for shape tracking. On top of the basic shape tracking algorithm, some simple track management algorithms have been realized.

Gating The signal processing algorithm described in Section 6.2 detects only points on the moving object. However, of course, the algorithm is not perfect and spurious measurements that do not stem from the moving object may arise. Clutter measurements may be caused by the sensor noise, light changes, or small movements of background objects. Nevertheless, if the signal processing parameters are adjusted well, clutter measurements are rather unlikely in the considered setting. Hence, the following very simple gating criterion for dismissing clutter is sufficient: Each measurement that lies in a scaled version of the last shape estimate is treated as object-generated, otherwise it is clutter.

Track Initialization and Termination Track initialization can also be performed in naïve manner as we restrict ourselves to a single extended object and the clutter rate is rather low. A new track can be initialized if the overall number of extracted measurements exceeds a user-defined value for some frames. The initial location of the track is chosen to be the mean of the measurements. As the shape of the object is a priori totally unknown, the first shape estimate (i.e., the parameter vector of the Fourier coefficients) is chosen to be a circle with a high uncertainty. Due to the simplifying assumptions in our scenario, a track can be terminated if the number of validated measurements falls below a user-defined value for a some frames.

6.4 Real-Time Implementation

This section briefly describes the details of a real-time implementation of the above suggested tracking system (see [5] and [30]).

The Microsoft Kinect sensor supplies RGB images with a resolution of 640x480 in 24 Bit and depth images with a resolution of 640x480 in 11 Bit, where 30 images per second are reached. Both the signal processing and shape tracking algorithm *UKF-SC-RHM* are executed on the same standard desktop computer (Intel Core 2 Quad-Core Q660 processor with 2.4 GHz, 8192 MB RAM, and AMD Radeon HD 5750 graphic card). The signal processing algorithm is written with the help of OpenCV [Bra00], OpenCL [Khr08] (for the optical flow algorithm) and OpenGL [Shr08]. The shape tracking algorithm is based on a C++ implementation of the *UKF* [JU04] that employs numerical functions of the Eigen library [GJ$^+$10].

With the above implementation, the maximum possible number of 30 frames per second is achieved for the evaluated scenarios (see the next section). Reasonably, the obtained frame rate is influenced by several different factors, e.g., the number of Fourier coefficients and the number of measurements per frame.

6.5 Evaluation

The evaluation of the proposed tracking system consists of two different parts:

- *Part 1:* In the first part, two extended object tracking algorithms that work on the extracted point measurements $\hat{\mathcal{y}}_k$ are compared (see Section 6.2). The first method is based on an *RHM* for star-convex shapes (*UKF-SC-RHM*) as described in Section 6.3. The second method is a variant of active contours [KWT88, BI98, JBU04] tailored to point measurements. The objective of this comparison is to highlight the benefits of *RHMs* for shape estimation based on point measurements.

- *Part 2:* As already mentioned earlier, the main purpose of the experiment is to evaluate extended object tracking algorithms. However, the overall tracking system, i.e., moving point detection in combination with the shape tracking algorithm based on a *RHM*, has a couple of benefits compared to standard object tracking algorithms. These advantages are elucidated in the second part of the evaluation.

For both parts, we consider two different object shapes, i.e., a "T"–shaped object and a "+"–shaped object. Of course, the object shape and the trajectory are a priori unknown to the tracking algorithm. Please note that experiments with further object types (different colors and shapes) can be found in [1], [5] and [30].

Part 1: RHM vs. Active Contours (based on Point Measurements)

In this section, the extracted point measurements from Section 6.2 are taken as measurements for the tracking algorithm. We perform a comparison of

- the shape tracking algorithm for star-convex shapes *UKF-SC-RHM* as described in Section 6.3 with

- a variant of active contours [KWT88, JBU04] tailored to point measurements.

Active contours are standard object tracking methods in computer vision. However, active contours are defined for intensity images, i.e., they do not directly work with point measurements. For this reason, the measurements $\hat{\mathcal{Y}}_k$ are interpreted as an intensity image by putting a Gaussian kernel at the locations of the measurements (see Fig. 6.3). In this vein, the adopted active contours calculate an enclosing contour of point measurements. The parameters of both algorithms have been tweaked to give the best results. Of course, the parameter adjusting has been performed only once per algorithm (for both shapes).

The shape estimates for both algorithms are depicted in Fig. 6.4 and Fig. 6.5. It can be observed that the *UKF-SC-RHM* gives precise and smooth shape estimates of the underlying train. The turn of the train is followed with a slight delay due to the underlying probabilistic system model and the few available measurements. The shape determined by the active contours are very imprecise and jitter. Because only a couple of measurements that do not completely cover the object's surface are available, active contours are not able to determine the contour precisely. Additionally, active contour models do not incorporate measurement noise, i.e., when there is a point not lying on the surface, the active contour model would probably enclose this outlier. *RHMs*, however, explicitly incorporate the measurement noise of each single measurement.

This scenario illustrates the need of an extended object tracking method that consists of an model for the generation of a *single* point measurement. It is not sufficient to calculate an enclosing shape of the measurements. The shape can only be estimated if the motion of the object is incorporated

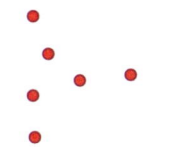 

(a) Example measurements.   (b) Active contour for "smoothed" measurements.

Figure 6.3: A point set can be interpreted as a continuous intensity image, i.e., a Gaussian mixture, by placing a Gaussian density at each point.

for several frames. At this point, it is essential to note that *RHMs* are not per se better than active contours because the primary application area of active contours are RGB images and not (sparse and noisy) point sets as in this experiment.

Part 2: Point Measurements + RHM vs. Active Contours on RGBD Data

The purpose of the second evaluation part is to demonstrate the benefits of the overall tracking system consisting of the moving point detection in combination with the *RHM* approach for shape estimation. Of course, there is huge variety of other tracking methods that are suitable for object tracking in RGB and/or depth data. However, when focusing on approaches, which are capable of estimating a detailed shape contour, the range of algorithms is rather limited. Here, we focus again on the popular active contour models because they can be seen as a standard contour tracking method.

Active Contours on RGB Image   First, it would be reasonable to apply active contours to the RGB images. However, this approach is unsuitable for the considered scenario. As the tracks have nearly the same color as the train, the contours get stuck in the tracks.

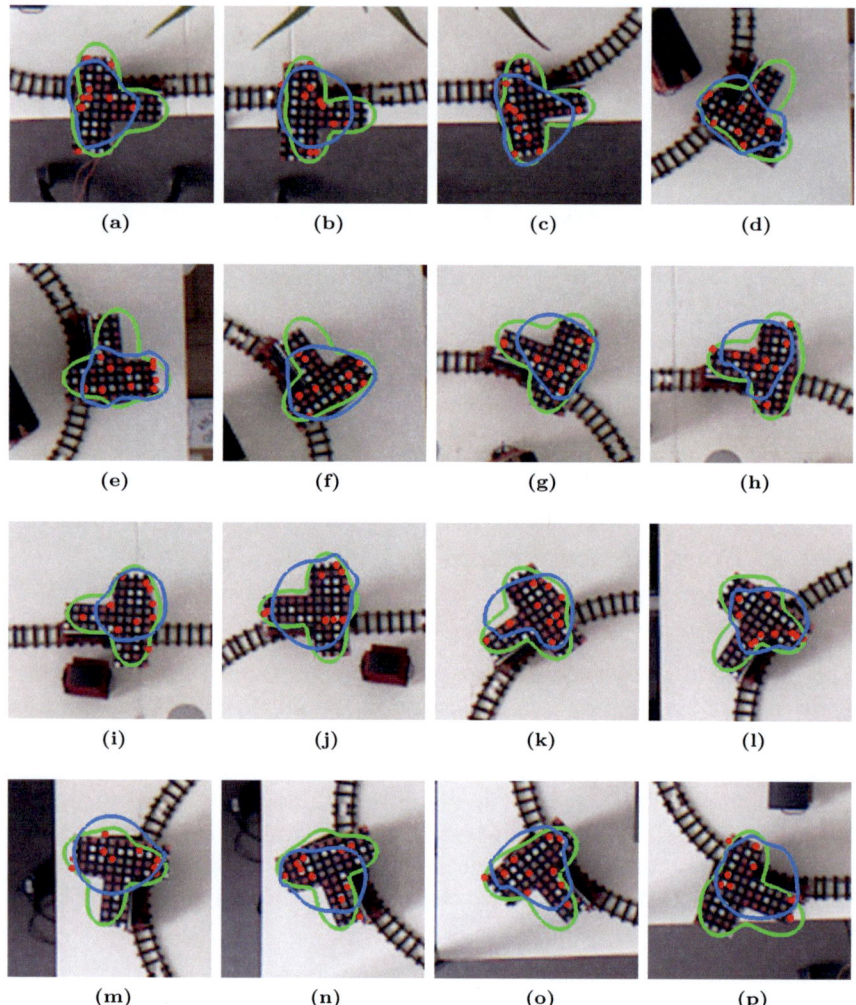

Figure 6.4: Tracking results for the "T"–shaped object: *RHM* (green) vs. active contour model modified for point measurements (blue).

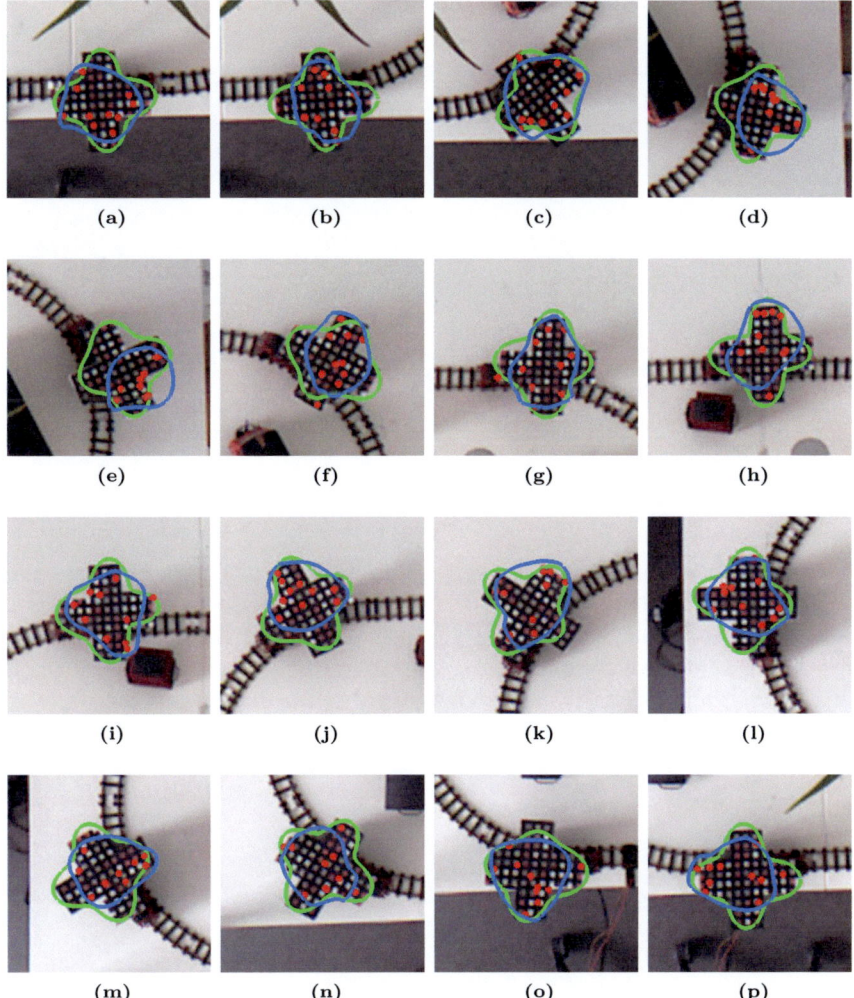

Figure 6.5: Tracking results for the "+"–shaped object: *RHM* (green) vs. active contour model modified for point measurements (blue).

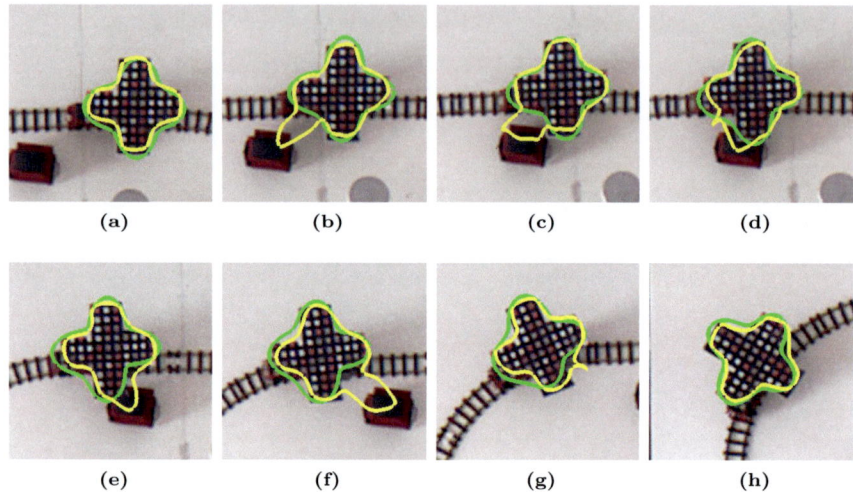

(a)   (b)   (c)   (d)

(e)   (f)   (g)   (h)

Figure 6.6: Tracking results for the "+"-shaped object: *RHM* (green) vs. active contour model applied to depth images (yellow).

Active Contours on Depth Image   A promising approach is to use the depth images instead of the RGB images. Actually, the active contour model for depth images works very well in the considered scenario in case the object significantly differs from the background in depth. However, if this is not the case, active contour may get stuck at non-moving objects, which belong to the background. Such a case is shown in Fig. 6.6, where the toy train passes a stationary background object that is located close to the tracks.

6.6   Conclusions

This chapter was about an example application in which a ground moving object is tracked with an RGB-D sensor from a bird's eye view. Although the focus was on a particular miniature setting, the characteristics and properties of the extracted point measurements are representative for many other relevant sensors such as *Ground Moving Target Indicator (GMTI)* radars. As the sensor is close to the target object, several measurements are resolved on its surface. However, due to the rather large sensor noise, only a couple of measurements that do not cover the entire surface are obtained per frame. As this is exactly an extended object problem as described in Chapter 2, we believe that the experiment can serve as an evaluation platform for extended object tracking algorithms.

With the help of the experiment, it was shown that an extended object problem cannot be solved naïvely by means of simple adoptions of standard methods. For this purpose, the popular active contour models have been modified to estimate an enclosing contour of point measurements. A comparison of the modified active contours with a star-convex *RHM* emphasized the necessity of a measurement model for a single point measurement (see Chapter 2).

The overall tracking system consisting of the moving point extraction algorithm plus the shape tracking algorithm based on an *RHM* is a promising alternative to other state-of-the-art tracking methods (in the considered scenario). It is suitable for many further applications such as people tracking, habitat monitoring, or traffic surveillance.

CHAPTER 7

Conclusions

Typically, tracking algorithms employ a point model of the target object, i.e., each received measurement is assumed to originate from a single point. For a so-called extended object, this assumption is not fulfilled anymore as several measurements from spatially distinct measurement sources on the object are obtained. Extended object tracking is a problem of growing interest as recent progress in sensor technology and new applications will render the incorporation of the object's extent in a tracking procedure inevitable in the future.

This thesis was devoted to the problem of simultaneously tracking the kinematic and shape parameters of a single extended object. As the shape of the object is unknown and part of the estimation problem, standard (linear) tracking algorithms cannot be applied anymore. Instead, a high-dimensional, nonlinear estimation problem has to be solved. Additionally, systematic errors may cause serious problems as the properties of the object's extent are usually unknown. Both two major challenges have been addressed in this thesis:

- The developed methods show that extended object tracking is feasible. Now, even complex shape information can be estimated from noisy point measurements with the help of a recursive Bayesian state estimator using closed-form formulas. This is a significant progress that not only leverages the application and value of extended object tracking but also opens new possibilities for higher-level information tasks.

  The fundamental approach that renders extended object tracking tractable is a novel model for extended objects called *Random Hypersurface Model* (*RHM*). The underlying idea of an *RHM* is to model the interior of a shape via scaling of the shape boundary. In this manner, the curve fitting techniques that have been developed in this thesis can be used for estimating region shapes.

- In this thesis, it was pointed out how systematic errors in the measurement sources can significantly bring down the estimation quality of a stochastic estimator, i.e., the estimates become inconsistent. In case the object's extent is known (or can be estimated from further information sources), it is possible to damp the negative effect of systematic errors. For this purpose, a so-called set-theoretic extent model was proposed that does not impose statistical assumptions on the measurement sources. Inference based on this model was performed with a so-called *Statistical and Set-theoretic Information (SSI) Filter* for which novel outer-bounding techniques have been developed.

Throughout the entire thesis, we employed Gaussian filters based on statistical linearization, which have significant advantages compared to other nonlinear filtering approaches, e.g., the likelihood does not have to be evaluated explicitly and closed-form expressions are often available. From a user's perspective, all developed methods are easy to implement and to integrate into other tracking frameworks. Essentially, only the developed measurement functions have to be fed into a standard nonlinear Gaussian filter and implementing the analytic approaches is trivial anyway. As all methods are based on Gaussians, it is not required to deal with sophisticated non-standard statistical methods.

The practical relevance and value of the theoretical results of this thesis have been demonstrated with an an illustrative example application, where the shape of a ground moving object is to be tracked with the help of RGB and depth data.

Outlook This thesis focused on a *single* extended object. Of course, many tracking systems must be able to deal with multiple objects from which some are extended and some not. There is a data association problem, i.e., it is not known which measurement originates from which object and clutter measurements that do not arise from any object can occur. In this context, elaborate mechanisms for track initialization and termination as well as splitting and merging of extended objects are required.

Reasonably, the description complexity of an object's shape should be as high as possible because a more realistic object model is expected to increase the tracking performance. However, it is essential to keep in mind that detailed shape information can only be obtained under particular circumstances, e.g., the measurement quality must be high enough and the object must not maneuver too fast. When these circumstances are not met, the uncertainty of the shape estimate may grow to infinity and the track is lost. As track loss is associated with a sequence of poor estimates and a loss of the object identity, it should be avoided at any prices. For this reason, there is a significant need for a mechanism that chooses the complexity of the used shape description reaching from a point model to general free-form shapes. How to choose the optimal description complexity is still an open question. In this context, performance bounds are required that say how precise the shape can be determined.

In this thesis, we restricted ourselves to noisy point measurements as they allow for modeling the most relevant sensors, e.g., laser rangefinders and radar devices. Nevertheless, it may be beneficial to work directly with nonlinear measurements models, e.g., bearing or distance measurements.

So far, we have used very simple motion models that assume that the shape parameters and kinematic parameters evolve independently. Of course, more elaborate motion models that allow for a coupling between the shape and kinematic parameters may be more suitable. Furthermore, it may be possible to estimate the evolution of the shape parameters, i.e., some kind of shape velocity vector, in addition to the shape parameters itself.

Future investigations may also focus on exploiting the shape information for further higher-level problems such as object classification, sensor management, and planning.

List of Figures

Own Publications

Journal Articles

[1] **Marcus Baum**. Student Research Highlight: Simultaneous Tracking and Shape Estimation of Extended Targets. *IEEE Aerospace and Electronic Systems Magazine*, 27(7):42–44, July 2012.

[2] **Marcus Baum** and Uwe D. Hanebeck. Extended Object Tracking Based on Set-Theoretic and Stochastic Fusion. *IEEE Transactions on Aerospace and Electronic Systems*, 48(4):3103–3115, October 2012.

[3] **Marcus Baum** and Uwe D. Hanebeck. Extended Object Tracking with Random Hypersurface Models (to appear). *IEEE Transactions on Aerospace and Electronic Systems*, accepted February 2013.

Conference Proceedings

[4] **Marcus Baum**, Florian Faion, and Uwe D. Hanebeck. Modeling the Target Extent with Multiplicative Noise. In *Proceedings of the 15th International Conference on Information Fusion (Fusion 2012)*, Singapore, July 2012.

[5] **Marcus Baum**, Florian Faion, and Uwe D. Hanebeck. Tracking Ground Moving Extended Objects using RGBD Data. In *Proceedings of the 2012 IEEE International Conference on Multisensor Fusion and Integration for Intelligent Systems (MFI 2012)*, Hamburg, Germany, September 2012.

[6] **Marcus Baum**, Michael Feldmann, Dietrich Fränken, Uwe D. Hanebeck, and Wolfgang Koch. Extended Object and Group Tracking: A Comparison of Random Matrices and Random Hypersurface Models. In *Proceedings of the IEEE ISIF Workshop on Sensor Data Fusion: Trends, Solutions, Applications (SDF 2010)*, Leipzig, Germany, October 2010.

[7] **Marcus Baum**, Ioana Gheta, Andrey Belkin, Jürgen Beyerer, and Uwe D. Hanebeck. Data Association in a World Model for Autonomous Systems. In *Proceedings of the 2010 IEEE International Conference on Multisensor Fusion and Integration for Intelligent Systems (MFI 2010)*, Salt Lake City, Utah, USA, September 2010.

[8] **Marcus Baum** and Uwe D. Hanebeck. Extended Object Tracking based on Combined Set-Theoretic and Stochastic Fusion. In *Proceedings of the 12th International Conference on Information Fusion (Fusion 2009)*, Seattle, Washington, USA, July 2009.

[9] **Marcus Baum** and Uwe D. Hanebeck. Random Hypersurface Models for Extended Object Tracking. In *Proceedings of the 9th IEEE International Symposium on Signal Processing and Information Technology (ISSPIT 2009)*, Ajman, United Arab Emirates, December 2009.

[10] **Marcus Baum** and Uwe D. Hanebeck. Tracking an Extended Object Modeled as an Axis-Aligned Rectangle. In *4th German Workshop on Sensor Data Fusion: Trends, Solutions, Applications (SDF 2009), 39th Annual Conference of the Gesellschaft für Informatik e.V. (GI)*, Lübeck, Germany, October 2009.

[11] **Marcus Baum** and Uwe D. Hanebeck. Association-free Tracking of Two Closely Spaced Targets. In *Proceedings of the 2010 IEEE International Conference on Multisensor Fusion and Integration for Intelligent Systems (MFI 2010)*, Salt Lake City, Utah, USA, September 2010.

[12] **Marcus Baum** and Uwe D. Hanebeck. Tracking a Minimum Bounding Rectangle based on Extreme Value Theory. In *Proceedings of the 2010 IEEE International Conference on Multisensor Fusion and Integration for Intelligent Systems (MFI 2010)*, Salt Lake City, Utah, USA, September 2010.

[13] **Marcus Baum** and Uwe D. Hanebeck. Fitting Conics to Noisy Data Using Stochastic Linearization. In *Proceedings of the 2011 IEEE/RSJ International Conference on Intelligent Robots and Systems (IROS 2011)*, San Francisco, California, USA, September 2011.

[14] **Marcus Baum** and Uwe D. Hanebeck. Shape Tracking of Extended Objects and Group Targets with Star-Convex RHMs. In *Proceedings*

of the 14th International Conference on Information Fusion (Fusion 2011), Chicago, Illinois, USA, July 2011.

[15] **Marcus Baum** and Uwe D. Hanebeck. Using Symmetric State Transformations for Multi-Target Tracking. In *Proceedings of the 14th International Conference on Information Fusion (Fusion 2011)*, Chicago, Illinois, USA, July 2011.

[16] **Marcus Baum**, Vesa Klumpp, and Uwe D. Hanebeck. A Novel Bayesian Method for Fitting a Circle to Noisy Points. In *Proceedings of the 13th International Conference on Information Fusion (Fusion 2010)*, Edinburgh, United Kingdom, July 2010.

[17] **Marcus Baum**, Benjamin Noack, Frederik Beutler, Dominik Itte, and Uwe D. Hanebeck. Optimal Gaussian Filtering for Polynomial Systems Applied to Association-free Multi-Target Tracking. In *Proceedings of the 14th International Conference on Information Fusion (Fusion 2011)*, Chicago, Illinois, USA, July 2011.

[18] **Marcus Baum**, Benjamin Noack, and Uwe D. Hanebeck. Extended Object and Group Tracking with Elliptic Random Hypersurface Models. In *Proceedings of the 13th International Conference on Information Fusion (Fusion 2010)*, Edinburgh, United Kingdom, July 2010.

[19] **Marcus Baum**, Benjamin Noack, and Uwe D. Hanebeck. Random Hypersurface Mixture Models for Tracking Multiple Extended Objects. In *Proceedings of the 50th IEEE Conference on Decision and Control (CDC 2011)*, Orlando, Florida, USA, December 2011.

[20] **Marcus Baum**, Patrick Ruoff, Dominik Itte, and Uwe D. Hanebeck. Optimal Point Estimates for Multi-Target States based on Kernel Distances. In *Proceedings of the 51st IEEE Conference on Decision and Control (CDC 2012)*, Maui, Hawaii, USA, December 2012.

[21] **Marcus Baum**, Peter Willett, and Uwe D. Hanebeck. Calculating Some Exact MMOSPA Estimates for Particle Distributions. In *Proceedings of the 15th International Conference on Information Fusion (Fusion 2012)*, Singapore, July 2012.

[22] Florian Faion, **Marcus Baum**, and Uwe D. Hanebeck. Tracking 3D Shapes in Noisy Point Clouds with Random Hypersurface Models.

In *Proceedings of the 15th International Conference on Information Fusion (Fusion 2012)*, Singapore, July 2012.

[23] Yvonne Fischer, **Marcus Baum**, Fabian Flohr, Uwe Hanebeck, and Jürgen Beyerer. Evaluation of Tracking Methods for Maritime Surveillance. In *Signal Processing, Sensor Fusion, and Target Recognition XXI (Proceedings of SPIE)*, Baltimore, Maryland, USA, April 2012.

[24] Ioana Gheta, **Marcus Baum**, Andrey Belkin, Jürgen Beyerer, and Uwe D. Hanebeck. Three Pillar Information Management System for Modeling the Environment of Autonomous Systems. In *Proceedings of the 2010 IEEE International Conference on Virtual Environments, Human-Computer Interfaces and Measurement Systems (VECIMS 2010)*, Taranto, Italy, September 2010.

[25] Reiner Hähnle, **Marcus Baum**, Richard Bubel, and Marcel Rothe. A Visual Interactive Debugger Based on Symbolic Execution. In *Proceedings of the 25th IEEE/ACM International Conference on Automated Software Engineering (ASE 2010)*, Antwerp, Belgium, September 2010.

[26] Vesa Klumpp, Benjamin Noack, **Marcus Baum**, and Uwe D. Hanebeck. Combined Set-Theoretic and Stochastic Estimation: A Comparison of the SSI and the CS Filter. In *Proceedings of the 13th International Conference on Information Fusion (Fusion 2010)*, Edinburgh, United Kingdom, July 2010.

[27] Benjamin Noack, **Marcus Baum**, and Uwe D. Hanebeck. Automatic Exploitation of Independencies for Covariance Bounding in Fully Decentralized Estimation. In *Proceedings of the 18th IFAC World Congress (IFAC 2011)*, Milan, Italy, August 2011.

[28] Benjamin Noack, **Marcus Baum**, and Uwe D. Hanebeck. Covariance Intersection in Nonlinear Estimation Based on Pseudo Gaussian Densities. In *Proceedings of the 14th International Conference on Information Fusion (Fusion 2011)*, Chicago, Illinois, USA, July 2011.

[29] Marc Reinhardt, Benjamin Noack, **Marcus Baum**, and Uwe D. Hanebeck. Analysis of Set-theoretic and Stochastic Models for Fusion under Unknown Correlations. In *Proceedings of the 14th International Conference on Information Fusion (Fusion 2011)*, Chicago, Illinois, USA, July 2011.

Supervised Student Works

[30] Kai Bouché. Implementierung eines echtzeitfähigen Trackingsystems basierend auf RGBD-Daten und sternkonvexen RHMs (Translation: Implementation of a Real-Time Tracking System based on RGBD Data and Star-Convex RHMs). Bachelor thesis, Intelligent-Sensor-Actuator Systems Laboratory, Karlsruhe Institute of Technology (KIT), 2012.

[31] Michel Conrad. Informationsgewinn in der Szenenbeschreibung (Translation: Information Gain in the Scene Model). Diploma thesis, Intelligent-Sensor-Actuator Systems Laboratory, Karlsruhe Institute of Technology (KIT), 2010.

[32] Florian Faion. Modellierung und Inferenz von unsicherem Wissen mit zufälligen Mengen (Translation: Modeling and Inference of Uncertain Knowledge with Random Sets). Diploma thesis, Intelligent-Sensor-Actuator Systems Laboratory, Karlsruhe Institute of Technology (KIT), 2010.

[33] Fabian Flohr. Implementierung und Evaluation echtzeitfähiger Multi-Target-Tracking-Verfahren in der NEST-Modellwelt (Translation: Implementation and Evaluation of Real-time Multi-Target Tracking Methods in the NEST Model). Master's thesis, Intelligent-Sensor-Actuator Systems Laboratory, Karlsruhe Institute of Technology (KIT), 2011.

[34] Dominik Itte. Assoziationsfreie Methoden zum Tracking von mehreren Objekten (Translation: Association-free Methods for Tracking Multiple Objects). Diploma thesis, Intelligent-Sensor-Actuator Systems Laboratory, Karlsruhe Institute of Technology (KIT), 2011.

[35] Marc Reinhardt. Fusion unsicherer Information unter unbekannten Korrelationen (Translation: Fusion of Uncertain Information under Unknown Correlation). Minor thesis, Intelligent Sensor-Actuator-Systems Laboratory, Karlsruhe Institute of Technology (KIT), 2011.

[36] Marc Reinhardt. Strategien zur Identifikation von Abhängigkeiten und zur dezentralen Informationsfusion in verteilten Sensornetzen (Translation: Strategies for the Identification of Dependencies for Decentralized Information Fusion in Distributed Sensor Networks). Diploma thesis, Intelligent Sensor-Actuator-Systems Laboratory, Karlsruhe Institute of Technology (KIT), 2011.

[37] Pascal Schirmer. Fusion von Sensordaten zur Echtzeitlokalisierung von Fahrzeugen (Translation: Fusion of Sensor Data for Real-Time Localization of Vehicles). Diploma thesis, Intelligent-Sensor-Actuator Systems Laboratory, Karlsruhe Institute of Technology (KIT), 2009.

Bibliography

[AMGC02]   M. Sanjeev Arulampalam, Simon Maskell, Neil Gordon, and Tim Clapp. A Tutorial on Particle Filters for On-line Nonlinear/Non-Gaussian Bayesian Tracking. *IEEE Transactions on Signal Processing*, 50:174–188, 2002.

[AOMMB07] Kai O. Arras, Óscar Martínez Mozos, and Wolfram Burgard. Using Boosted Features for the Detection of People in 2D Range Data. In *Proc. IEEE International Conference on Robotics and Automation (ICRA'07)*, Rome, Italy, 2007.

[ARR99]   Sung Joon Ahn, Wolfgang Rauh, and Matthias Recknagel. Ellipse Fitting and Parameter Assessment of Circular Object Targets for Robot Vision. In *Proc. IEEE/RSJ Int. Conf. Intelligent Robots and Systems IROS '99*, volume 1, pages 525–530, 1999.

[ASC09]   Ali Al-Sharadqah and Nikolai Chernov. Error Analysis for Circle Fitting Algorithms. *Electronic Journal of Statistics*, 3:886–911, 2009.

[BC91]   Ted J. Broida and Rama Chellappa. Estimating the Kinematics and Structure of a Rigid Object from a Sequence of Monocular Images. *IEEE Transactions on Pattern Analysis and Machine Intelligence*, 13(6):497–513, June 1991.

[BDT+06]   Yvo Boers, Hans Driessen, Johan Torstensson, Mikael Trieb, Rickard Karlsson, and Fredrik Gustafsson. Track-Before-Detect Algorithm for Tracking Extended Targets. *IEE Proceedings on Radar, Sonar and Navigation*, 153(4):345–351, August 2006.

[BHH10]   Frederik Beutler, Marco F. Huber, and Uwe D. Hanebeck. Optimal Stochastic Linearization for Range-based Localization. In *Proceedings of the 2010 IEEE/RSJ International Conference on Intelligent Robots and Systems (IROS 2010)*, Taipei, Taiwan, October 2010.

[BI98]   Andrew Blake and Michael Isard. *Active Contours.* Springer-Verlag New York, Inc., Secaucus, NJ, USA, 1st edition, 1998.

[Bra00]   Gary Bradski. The OpenCV Library. *Dr. Dobb's Journal of Software Tools,* 2000.

[Bro11]   Mike Brookes. *The Matrix Reference Manual,* 2011. [online] http://www.ee.imperial.ac.uk/hp/staff/dmb.

[BSDH09]   Yaakov. Bar-Shalom, Fred Daum, and Jim Huang. The Probabilistic Data Association Filter. *IEEE Control Systems Magazine,* 29(6):82–100, December 2009.

[BSKL02]   Yaakov Bar-Shalom, Thiagalingam Kirubarajan, and X.-Rong Li. *Estimation with Applications to Tracking and Navigation.* John Wiley & Sons, Inc., New York, NY, USA, 2002.

[BSWT11]   Yaakov Bar-Shalom, Peter K. Willett, and Xin Tian. *Tracking and Data Fusion: A Handbook of Algorithms.* YBS Publishing, 2011.

[CGW⁺09]   David F. Crouse, Marco Guerriero, Peter Willett, Roy Streit, and Darin Dunham. A Look at the PMHT. In *Proceedings of the 12th International Conference on Information Fusion (Fusion 2009),* July 2009.

[Cha65]   Nai Ng Chan. On Circular Functional Relationships. *Journal of the Royal Statistical Society. Series B (Methodological),* 27(1):45–56, 1965.

[Che10]   Nikolai Chernov. *Circular and Linear Regression: Fitting Circles and Lines by Least Squares.* CRC Press, 2010.

[CKS95]   Vicent Caselles, Ron Kimmel, and Guillermo Sapiro. Geodesic Active Contours. *International Journal of Computer Vision,* 22:61–79, 1995.

[CL05]   Nikolai Chernov and Claire Lesort. Least Squares Fitting of Circles. *Journal of Mathematical Imaging and Vision,* 23(3):239–252, 2005.

[COY10]   Alexander Carballo, Akihisa Ohya, and Shin'ichi Yuta. People Detection using Range and Intensity Data from Multi-layered Laser Range Finders. In *Proceedings of the IEEE/RSJ International Conference on Intelligent Robots and Systems (IROS)*, pages 5849–5854, 2010.

[CRM03]   Dorin Comaniciu, Visvanathan Ramesh, and Peter Meer. Kernel-based Object Tracking. *IEEE Transactions on Pattern Analysis and Machine Intelligence*, pages 564–577, 2003.

[CRSC06]   Raymond J. Carroll, David Ruppert, Leonard A. Stefanski, and Ciprian M. Crainiceanu. *Measurement Error in Nonlinear Models: A Modern Perspective, Second Edition*. Chapman & Hall/CRC, June 2006.

[CSG12]   Avishy Carmi, François Septier, and Simon J. Godsill. The Gaussian Mixture MCMC Particle Algorithm for Dynamic Cluster Tracking. *Automatica*, 48(10):2454–2467, 2012.

[DN99]   Joachim Denzler and Heinrich Niemann. Active Rays: Polar-Transformed Active Contours for Real-Time Contour Tracking. *Real-Time Imaging*, 5:203–213, June 1999.

[EAB92]   Tim Ellis, Ahmed Abbood, and Beatrice Brillault. Ellipse Detection and Matching with Uncertainty. *Image and Vision Computings*, 10(5):271–276, 1992.

[FB08]   Iuri Frosio and N. Alberto Borghese. Real-time Accurate Circle Fitting with Occlusions. *Pattern Recognition*, 41(3):1041–1055, 2008.

[FF08]   Michael Feldmann and Dietrich Fränken. Tracking of Extended Objects and Group Targets using Random Matrices – A New Approach. *Proceedings of the 11th International Conference on Information Fusion (Fusion 2008)*, July 2008.

[FF09]   Michael Feldmann and Dietrich Fränken. Advances on Tracking of Extended Objects and Group Targets using Random Matrices. In *Proceedings of the 12th International Conference on Information Fusion (Fusion 2009)*, Seattle, Washington, July 2009.

[FFK11]   Michael Feldmann, Dietrich Fränken, and Wolfgang Koch. Tracking of Extended Objects and Group Targets using Random Matrices. *IEEE Transactions on Signal Processing*, 59(4):1409–1420, 2011.

[FK08]   Michael Feldmann and Wolfgang Koch. Road-map Assisted Convoy Track Maintenance using Random Matrices. *11th International Conference on Information Fusion*, pages 1–8, July 2008.

[Frä07]   Dietrich Fränken. Consistent Unbiased Linear Filtering with Polar Measurements. In *Proceedings of the 10th International Conference on Information Fusion (Fusion 2007)*, July 2007.

[GBCR00]   Xiang Gao, T.E. Boult, Frans Coetzee, and Visvanathan Ramesh. Error Analysis of Background Adaption. In *Proceedings of the IEEE Conference on Computer Vision and Pattern Recognition*, volume 1, pages 503–510, 2000.

[GBL+11]   Nicola Greggio, Alexandre Bernardino, Cecilia Laschi, José Santos-Victor, and Paolo Dario. Real-Time 3D Stereo Tracking and Localizing of Spherical Objects with the iCub Robotic Platform. *Journal of Intelligent & Robotic Systems*, 2011.

[GGMS05]   Kevin. Gilholm, Simon Godsill, Simon Maskell, and David Salmond. Poisson Models for Extended Target and Group Tracking. In *SPIE: Signal and Data Processing of Small Targets*, 2005.

[GJ+10]   Gaël Guennebaud, Benoît Jacob, et al. Eigen v3. http://eigen.tuxfamily.org, 2010.

[GLO10]   Karl Granström, Christian Lundquist, and Umut Orguner. A Gaussian Mixture PHD filter for Extended Target Tracking. In *Proceedings of the 13th International Conference on Information Fusion (Fusion 2010)*, Edinburgh, Scotland, July 2010.

[GML+08]   Nicola Greggio, Luigi Manfredi, Cecilia Laschi, Paolo Dario, and Maria Chiara Carrozza. RobotCub Implementation of

Real-Time Least-Square Fitting of Ellipses. In *Proceedings of the 8th IEEE-RAS International Conference on Humanoid Robots (Humanoids 2008)*, pages 174–181, 2008.

[GO12]   Karl Granström and Umut Orguner. A PHD Filter for Tracking Multiple Extended Targets Using Random Matrices. *IEEE Transactions on Signal Processing*, 60(11):5657–5671, November 2012.

[GS05]   Kevin Gilholm and David Salmond. Spatial Distribution Model for Tracking Extended Objects. *IEE Proceedings on Radar, Sonar and Navigation*, 152(5):364–371, October 2005.

[GSDB07]   Joakim Gunnarsson, Lennart Svensson, Lars Danielsson, and Fredrik Bengtsson. Tracking Vehicles using Radar Detections. In *IEEE Intelligent Vehicles Symposium*, pages 296–302, June 2007.

[HH99a]   Uwe D. Hanebeck and Joachim Horn. A New Concept for State Estimation in the Presence of Stochastic and Set Theoretic Uncertainties. In *Proceedings of the 1999 IEEE/RSJ International Conference on Intelligent Robots and Systems (IROS 1999)*, Kjongju, Republic of Korea, October 1999.

[HH99b]   Uwe D. Hanebeck and Joachim Horn. A New Estimator for Mixed Stochastic and Set Theoretic Uncertainty Models Applied to Mobile Robot Localization. In *Proceedings of the 1999 IEEE International Conference on Robotics and Automation (ICRA 1999)*, pages 1335–1340, Detroit, Michigan, USA, May 1999.

[HH99c]   Uwe D. Hanebeck and Joachim Horn. A New State Estimator for a Mixed Stochastic and Set Theoretic Uncertainty Model. In *Proceedings of SPIE, Vol. 3720, AeroSense Symposium*, pages 336–344, Orlando, Florida, USA, April 1999.

[HH99d]   Uwe D. Hanebeck and Joachim Horn. New Estimators for Mixed Stochastic and Set Theoretic Uncertainty Models: The Scalar Measurement Case. In *Proceedings of the 1999*

IEEE Conference on Decision and Control (CDC 1999), pages 1934–1939, Phoenix, Arizona, USA, December 1999.

[HH08] Marco F. Huber and Uwe D. Hanebeck. Gaussian Filter based on Deterministic Sampling for High Quality Nonlinear Estimation. In *Proceedings of the 17th IFAC World Congress (IFAC 2008)*, volume 17, Seoul, Republic of Korea, July 2008.

[HHS99] Uwe D. Hanebeck, Joachim Horn, and Günther Schmidt. On Combining Statistical and Set Theoretic Estimation. *Automatica*, 35(6):1101–1109, June 1999.

[HS81] Berthold K.P. Horn and Brian G. Schunck. Determining Optical Flow. *Artificial Intelligence*, 17(1-3):185–203, 1981.

[IG03] Norikazu Ikoma and Simon Godsill. Extended Object Tracking with Unknown Association, Missing Observations, and Clutter using Particle Filters. In *IEEE Workshop on Statistical Signal Processing*, pages 502–505, 2003.

[Iss18] Leon Isserlis. On A Formula For The Product-Moment Coefficient of Any Order of a Normal Frequency Distribution in Any Number of Variables. *Biometrika*, 12(1-2):134–139, 1918.

[JBC08] M.H. Jaward, David Bull, and Nishan C. Canagarajah. Contour Tracking of Contaminant Clouds with Sequential Monte Carlo Methods. In *Proceedings of the IEEEInternational Conference on Acoustics, Speech and Signal Processing (ICASSP 2008)*, pages 1469–1472, April 2008.

[JBC10] M.H. Jaward, David Bull, and Nishan C. Canagarajah. Sequential Monte Carlo Methods for Contour Tracking of Contaminant Clouds. *Signal Processing*, 90:249–260, January 2010.

[JBR06] Ronald Jones, David M. Booth, and Nicholas J. Redding. Video Moving Target Indication in the Analysts' Detection Support System. Technical Report DSTO-RR-0306, DSTO, Edinburgh, S. Aust., 2006.

[JBU04]   Mathews Jacob, Thierry Blu, and Michael Unser. Efficient energies and algorithms for parametric snakes. *IEEE Transactions on Image Processing*, pages 1231–1244, 2004.

[JKS95]   Ramesh Jain, Rangachar Kasturi, and Brian G. Schunck. *Machine Vision*. McGraw-Hill, Inc., New York, NY, USA, 1995.

[JU04]   Simon J. Julier and Jeffrey K. Uhlmann. Unscented Filtering and Nonlinear Estimation. In *Proceedings of the IEEE*, volume 92, pages 401–422, 2004.

[Kan96]   Kenichi Kanatani. *Statistical Optimization for Geometric Computation: Theory and Practice*. Elsevier Science Inc., New York, NY, USA, 1996.

[Kan08]   Raymond Kan. From Moments of Sum to Moments of Product. *Journal of Multivariate Analysis*, 99(3):542–554, March 2008.

[Kar91]   Veikko Karimäki. Effective Circle Fitting for Particle Trajectories. *Nuclear Instruments and Methods in Physics Research A*, 305:187–191, July 1991.

[KD03]   J.H. Kotecha and P.M. Djuric. Gaussian Particle Filtering. *IEEE Transactions on Signal Processing*, 51(10):2592 – 2601, Oct. 2003.

[Khr08]   Khronos OpenCL Working Group. *The OpenCL Specification, version 1.0.29*, 8 December 2008.

[KKU06]   Wolfgang Koch, Jost Koller, and Martin Ulmke. Ground Target Tracking and Road Map Extraction. *ISPRS Journal of Photogrammetry and Remote Sensing*, 61(3–4):197–208, 2006.

[Koc08]   Wolfgang Koch. Bayesian Approach to Extended Object and Cluster Tracking using Random Matrices. *IEEE Transactions on Aerospace and Electronic Systems*, 44(3):1042–1059, July 2008.

[KS05]   Wolfgang Koch and Roman Saul. A Bayesian Approach to Extended Object Tracking and Tracking of Loosely Structured Target Groups. In *Proceedings of the 8th International Conference on Information Fusion (Fusion 2005)*, volume 1, July 2005.

[KWT88]   Michael Kass, Andrew Witkin, and Demetri Terzopoulos. Snakes: Active contour models. *International Journal of Computer Vision*, 1(4):321–331, 1988.

[LAB09]   Boris Lau, Kai Arras, and Wolfram Burgard. Tracking Groups of People with a Multi-Model Hypothesis Tracker. In *Proceedings of the IEEE International Conference on Robotics and Automation (ICRA 2009)*, pages 3180–3185, May 2009.

[LABS06]   Mihai Bogdan Luca, Stéphane Azou, Gilles Burel, and Alexandru Serbanescu. On Exact Kalman Filtering of Polynomial Systems. *IEEE Transactions on Circuits and Systems I: Regular Papers*, 53(6):1329 –1340, 2006.

[LBDS04]   Tine Lefebvre, Herman Bruyninckx, and Joris De Schutter. Kalman Filters for Non-linear Systems: A Comparison of Performance. *International Journal of Control*, 77(7):639–653, 2004.

[LOG11]   Christian Lundquist, Umut Orguner, and Fredrik Gustafsson. Extended Target Tracking Using Polynomials With Applications to Road-Map Estimation. *IEEE Transactions on Signal Processing*, 59(1):15–26, January 2011.

[LRL12a]   Jian Lan and X. Rong Li. Tracking of Extended Object or Target Group Using Random Matrix - Part I: New Model and Approach. In *Proceedings of the 15th International Conference on Information Fusion (Fusion 2012)*, pages 2177–2184, July 2012.

[LRL12b]   Jian Lan and X. Rong Li. Tracking of Extended Object or Target Group Using Random Matrix - Part II: Irregular Object. In *Proceedings of the 15th International Conference on Information Fusion (Fusion 2012)*, pages 2185–2192, July 2012.

[Mah07]   Ronald P. S. Mahler. *Statistical Multisource-Multitarget Information Fusion.* Artech House, Boston, Mass., 2007.

[Mah09]   Ronald P. S. Mahler. PHD Filters for Nonstandard Targets, I: Extended Targets. In *Proceedings of the 12th International Conference on Information Fusion (Fusion 2009)*, Seattle, Washington, July 2009.

[MS91]   Darryl R. Morrell and Wynn C. Stirling. Set-valued Filtering and Smoothing. *IEEE Transactions on Systems, Man and Cybernetics*, 21(1):184–193, 1991.

[MS03]   Darryl R. Morrell and Wynn C. Stirling. An Extended Set-valued Kalman Filter. In *ISIPTA '03*, pages 395–405, 2003.

[NKH09]   Benjamin Noack, Vesa Klumpp, and Uwe D. Hanebeck. State Estimation with Sets of Densities considering Stochastic and Systematic Errors. In *Proceedings of the 12th International Conference on Information Fusion (Fusion 2009)*, Seattle, Washington, USA, July 2009.

[NR97]   Garry N. Newsam and Nicholas J. Redding. Fitting the Most Probable Curve to Noisy Observations. In *Proceedings of the International Conference on Image Processing*, volume 2, pages 752–755, October 1997.

[NVMBS08]   Pedro Nuez, Ricardo Vazquez-Martin, Antonio Bandera, and Francisco Sandoval. An Algorithm for Fitting 2-D Data on the Circle: Applications to Mobile Robotics. *IEEE Signal Processing Letters*, 15:127–130, 2008.

[OGL11]   Umut Orguner, Karl Granstrom, and Christian Lundquist. Extended Target Tracking with a Cardinalized Probability Hypothesis Density Filter. In *Proceedings of the 14th International Conference on Information Fusion (Fusion 2011)*, Chicago, Illinois, USA, July 2011.

[Org12]   Umut Orguner. A Variational Measurement Update for Extended Target Tracking With Random Matrices. *IEEE Transactions on Signal Processing*, 60(7):3827 –3834, July 2012.

[PD11]   Benjamin Pannetier and Jean Dezert. Extended and Multiple Target Tracking: Evaluation of an Hybridization Solution. In *Proceedings of the 14th International Conference on Information Fusion (Fusion 2011)*, July 2011.

[PGMA12]   Nikolay Petrov, Amadou Gning, Lyudmila Mihaylova, and Donka Angelova. Box Particle Filtering for Extended Object Tracking. In *Proceedings of the 15th International Conference on Information Fusion (Fusion 2012)*, pages 82–89. IEEE, July 2012.

[PMGA11]   Nikolay Petrov, Lyudmila Mihaylova, Amadou Gning, and Donka Angelova. A Novel Sequential Monte Carlo Approach for Extended Object Tracking Based on Border Parameterisation. In *Proceedings of the 14th International Conference on Information Fusion (Fusion 2011)*, Chicago, Illinois, USA, July 2011.

[PMGA12]   Nikolay Petrov, Lyudmila Mihaylova, Amadou Gning, and Donka Angelova. Group Object Tracking with a Sequential Monte Carlo Method Based on a Parameterised Likelihood Function. *Monte Carlo Methods and Applications*, 2012.

[Por90]   John Porrill. Fitting Ellipses and Predicting Confidence Envelopes Using a Bias Corrected Kalman Filter. *Image Vision Computing*, 8:37–41, 1990.

[PP02]   Athanasios Papoulis and S. Unnikrishna Pillai. *Probability, Random Variables and Stochastic Processes*. McGraw-Hill, 4 edition, 2002.

[PP08]   Kaare Brandt Petersen and Michael Syskind Pedersen. The Matrix Cookbook, 2008.

[RAG04]   Branko Ristic, S. Arulampalam, and Neil Gordon. *Beyond the Kalman filter: Particle filters for tracking applications*. Artech House Publishers, 2004.

[Ros93]   Paul L. Rosin. A Note on the Least Squares Fitting of Ellipses. *Pattern Recognition Letters*, 14(10):799 – 808, 1993.

[SC10]   Anthony Swain and Daniel Clark. Extended Object Filtering using Spatial Independent Cluster Processes. In *Proceedings of the 13th International Conference on Information Fusion (Fusion 2010)*, Edinburgh, United Kingdom, July 2010.

[SCG09]   Francois Septier, Avishy Carmi, and Simon Godsill. Tracking of Multiple Contaminant Clouds. In *Proceedings of the 12th International Conference on Information Fusion (Fusion 2009)*, pages 1280–1287, July 2009.

[Shr08]   Dave Shreiner. *OpenGL Programming Guide : The Official Guide To Learning OpenGL, Version 2.1, 6/E*. Addison-Wesley Professional, August 2008.

[Sim06]   Dan Simon. *Optimal State Estimation: Kalman, H Infinity, and Nonlinear Approaches*. Wiley & Sons, 1. edition, August 2006.

[SSB+07]   Kleydis V. Suarez, Jesus C. Silva, Yannik Berthoumieu, Pedro. Gomis, and Mohamed Najim. ECG Beat Detection Using a Geometrical Matching Approach. *IEEE Transactions on Biomedical Engineering*, 54(4):641–650, April 2007.

[VIG04]   Jaco Vermaak, Norikazu Ikoma, and Simon Godsill. Extended object tracking using particle techniques. In *Proceedings of the Aerospace Conference*, volume 3, March 2004.

[VIG05]   Jaco Vermaak, Norikazu Ikoma, and Simon Godsill. Sequential Monte Carlo Framework for Extended Object Tracking. *IEE Proceedings on Radar, Sonar and Navigation*, 152(5):353–363, October 2005.

[WD04]   Milton J. Waxman and Oliver E. Drummond. A Bibliography of Cluster (Group) Tracking. *Signal and Data Processing of Small Targets 2004*, 5428(1):551–560, 2004.

[Wei]   Eric W. Weisstein. Circle - Circle Intersection. From MathWorld – A Wolfram Web Resource available online at http://mathworld.wolfram.com/Circle-CircleIntersection.html.

[WK01]   Michael Werman and Daniel Keren. A Bayesian Method
            for Fitting Parametric and Nonparametric Models to Noisy
            Data. *IEEE Transactions on Pattern Analysis and Machine
            Intelligence*, 23(5):528–534, 2001.

[WK10]   Monika Wieneke and Wolfgang Koch. Probabilistic Track-
            ing of Multiple Extended Targets using Random Matrices.
            In *Proceedings of the SPIE: Signal and Data Processing of
            Small Targets*, volume 7698, April 2010.

[WK12]   M. Wieneke and W. Koch. A PMHT Approach for Ex-
            tended Objects and Object Groups. *IEEE Transactions
            on Aerospace and Electronic Systems*, 48(3):2349–2370, July
            2012.

[YJS06]   Alper Yilmaz, Omar Javed, and Mubarak Shah. Object
            Tracking: A Survey. *ACM Computing Surveys*, 38(4),
            December 2006.

[YP05]   Masayuki Yokoyama and Tomaso Poggio. A Contour-Based
            Moving Object Detection and Tracking. In *Proceedings of
            the 14th International Conference on Computer Commu-
            nications and Networks*, pages 271–276, Washington, DC,
            USA, 2005. IEEE Computer Society.

[YSZ+11]   Hanxuan Yang, Ling Shao, Feng Zheng, Liang Wang, and
            Zhan Song. Recent Advances and Trends in Visual Tracking:
            A Review. *Neurocomputing*, 74(18):3823–3831, November
            2011.

[Zha97]   Zhengyou Zhang. Parameter Estimation Techniques: A Tu-
            torial with Application to Conic Fitting. *Image and Vision
            Computing*, 15(1):59–76, 1997.

[ZL05]   Dengsheng Zhang and Guojun Lu. Study and Evaluation of
            Different Fourier Methods for Image Retrieval. *Image and
            Vision Computing*, 23(1):33 – 49, 2005.

[ZXA06]   Sen Zhang, Lihua Xie, and Martin David Adams. Feature
            Extraction for Outdoor Mobile Robot Navigation based on
            a Modified Gauss-Newton Optimization Approach. *Robotics
            and Autonomous Systems*, 54(4):277–287, 2006.

Karlsruhe Series on Intelligent Sensor-Actuator-Systems

Edited by Prof. Dr.-Ing. Uwe D. Hanebeck // ISSN 1867-3813

Die Bände sind unter www.ksp.kit.edu als PDF frei verfügbar oder als Druckausgabe bestellbar.

Karlsruhe Series on
Intelligent Sensor-Actuator-Systems

Edited by Prof. Dr.-Ing. Uwe D. Hanebeck // ISSN 1867-3813

Die Bände sind unter www.ksp.kit.edu als PDF frei verfügbar oder als Druckausgabe bestellbar.